T0201194

Stress Management
in the Construction Industry

Stress Management in the Construction Industry

Mei-yung Leung

Department of Civil & Architectural Engineering
City University of Hong Kong

Isabelle Yee Shan Chan

Department of Real Estate and Construction
University of Hong Kong

Cary L. Cooper

Lancaster University Management School
Lancaster University
UK

WILEY Blackwell

This edition first published 2015
© 2015 by John Wiley & Sons, Ltd

Registered Office
John Wiley & Sons, Ltd, The Atrium, Southern Gate, Chichester, West Sussex, PO19 8SQ, United Kingdom.

Editorial Offices
9600 Garsington Road, Oxford, OX4 2DQ, United Kingdom.
The Atrium, Southern Gate, Chichester, West Sussex, PO19 8SQ, United Kingdom.

For details of our global editorial offices, for customer services and for information about how to apply for permission to reuse the copyright material in this book please see our website at www.wiley.com/wiley-blackwell.

Library of Congress Cataloging-in-Publication Data

Leung, Mei-yung.
  Stress management in the construction industry / Mei-yung Leung, Isabelle Yee Shan Chan, Cary Cooper.
      pages cm
  Includes bibliographical references and index.
  ISBN 978-1-118-45641-5 (cloth)
1. Construction industry–Employees–Health and hygiene.   2. Construction workers–Job stress.
3. Stress management.   I. Title.
  HD8039.B9L48 2015
  624.068'3–dc23
                                                    2014030548

A catalogue record for this book is available from the British Library.

Wiley also publishes its books in a variety of electronic formats. Some content that appears in print may not be available in electronic books.

Cover image courtesy of iStock Photo

Set in 10/12pt Sabon by SPi Publisher Services, Pondicherry, India
Printed and bound in Malaysia by Vivar Printing Sdn Bhd

1   2015

Contents

About the Authors

Dr Mei-yung Leung is Associate Professor at the City University of Hong Kong, PRC. She has more than 20 years of practical/teaching experience in the construction industry/education and has participated in a number of prestigious construction projects in Hong Kong. She has over 150 international publications in various research areas covering stress management, construction project management, value management, facility management and construction education. Dr Leung has successfully completed all levels of the Mindfulness-based Stress Reduction training at the University of Massachusetts in the United States, and conducted various stress management seminars and full courses to construction companies, professional institutes, higher educations and religious organisations. She has received a number of international awards, including the Tony Toy Memorial Award issued by the Hong Kong Institute of Value Management in Hong Kong, the Thomas D. Snodgrass Value Teaching Award issued by the SAVE International 'The Value Society' in the United States, and the Teaching Excellence Award issued by the City University of Hong Kong. Dr Leung is also a senior Fulbright Scholar at the Pennsylvania State University and University of Southern California.

Dr Isabelle Yee Shan Chan is Lecturer at the University of Hong Kong, PRC. She is the author of over 40 international publications in books, journals and conferences, covering areas of stress management, health and safety, culture and innovation in construction. In line with these research areas, she has participated in more than 10 research projects in the capacities of principal investigator, co-investigator and project coordinator. Stress management in construction is the research area of her PhD study; she has also successfully completed the mindfulness-based stress reduction programme in the Hospital Authority in Hong Kong. Dr Chan is the vice-chairman of the Institute of Safety and Health Practitioners in Hong Kong, and is also a visiting fellow of Hughes Hall, University of Cambridge, UK.

Professor Sir Cary Cooper is Professor of Organisational Psychology and Health, Lancaster University Management School, UK. He is the author of over 120 books (on occupational stress, women at work and industrial and organisational psychology), has written over 400 scholarly articles, and is a frequent contributor to national newspapers, TV and radio. Professor Cooper is a Fellow of the British Academy of Management and also of the Academy of Management (having also won the 1998 Distinguished Service

Award). In 2001 he was awarded a CBE in the Queen's Birthday Honours List for his contribution to organisational health, and in 2014 he was awarded a knighthood. Professor Cooper was the lead scientist to the UK Government Office for Science on their Foresight programme on Mental Capital and Well Being (2007–2008). He was appointed a member of the expert group on establishing guidance for the National Institute for Health and Clinical Excellence on 'promoting mental well-being through productive and healthy working conditions', 2009. Professor Cooper is Chair of the UK's Academy of Social Sciences (an umbrella body of 47 learned societies in the social sciences); and was Chair of the Chronic Disease and Well-being Global Agenda Council of the World Economic Forum in Geneva in 2009.

Preface

Due to the task complexity, tight timeframes, complicated work relationships, poor working environments and other factors, the construction industry has long been recognised as a stressful one. A survey study conducted by the Chartered Institute of Building (2006) indicates that nearly 70% of construction personnel have suffered from stress, anxiety or depression directly resulting from their work. Stress is not only a matter for the individual, but is also a real cost for any project, organisation, industry, and even nation (e.g., due to work-related stress, more than 10 million working days were lost in the UK; Health and Safety Executive 2012). To survive in such a demanding and dynamic industry, with these numerous sources of stress, construction personnel must be able to adopt suitable coping behaviours. However, not all coping behaviours have a positive effect on the individual. Studies indicate that the adoption of maladaptive coping is not unusual among construction personnel. In addition, although health care is now receiving increasing attention from the construction industry, the majority of training events and guidelines in this area only address physical well-being and neglect the importance of psychological health.

Hence, this book aims to enhance the performance of construction personnel by developing and presenting an integrated and comprehensive stress management model. This will not only illustrate how construction personnel are affected by various stressors and how this influences their performance, but will also explain how stress levels can be managed by dealing with the various stressors and using appropriate adaptive coping behaviours. To achieve this aim, this book, based on an extensive literature review, survey studies and scenario analyses, investigates various components of stress management for construction personnel, including the multiple dimensions of the stress they experience, the nature of their stressors, the coping behaviours they adopt, and the consequences of stress for their performance.

Chapter 1 sets out the aims and objectives of this book and provides an overview of the characteristics of the construction industry and the current state of research in this area. It summarises the background information, both practical and academic, required to contextualise the discussion which follows.

Chapter 2 introduces the historical development of stress theories since the 1920s. The concept of stress was first developed by a group of psychobiologists and then extended by other psychobiologists, sociologists and psychiatrists in different contexts. Various stress theories (mainly arousal and appraisal and regulatory theories) and their implications are discussed.

Chapter 3 discusses the stress experienced by construction personnel and explains the arousal mechanisms for different types of stress by reference to different theories. Three main kinds of stress (work stress arising from person–environment misfit, physical stress from the homeostatic effect, and emotional stress governed by the limbic system) are investigated, based on both quantitative and qualitative studies of construction professionals. Some practical implications are then set out at the end of the chapter.

Chapter 4 introduces the sources of stress (i.e., stressors) in detail. Five main types including personal, interpersonal, task, organisational and physical are identified, and the results of a scientific investigation into how different types of stressors induce the three main kinds of stress among different construction personnel are discussed. Based on these findings, recommendations are made about the identification and assessment of key stressors and how these can be reduced for construction staff.

Chapter 5 analyses the consequences of stress (i.e., performance) in terms of stress management in construction projects. Performance is classified along various dimensions, including personal, task and organisational, and each type is described in detail in the construction industry context. Conceptual models of stress and performance are then developed to illustrate the impact of stress on different types of performance for construction personnel. The results of a questionnaire survey and case study analysis are presented to show the significant relationship between stress and performance. These findings are then used as a basis for recommendations for how to manage the stress levels of construction personnel and optimise their ultimate project performance.

Chapter 6 deals with the coping behaviours adopted by construction personnel. Three main types are addressed: problem-, emotion- and meaning-based coping. Scientific research shows that different forms of coping behaviours have different effects on performance. To better manage stress among construction personnel, several practical implications from this body of work are presented.

Chapter 7 concludes the book and sets out an integrated stressor–stress–coping behaviour–performance model, ending with a summary of its recommendations.

We hope that this book will prove useful to both academics and practitioners. For the former, the empirical support to the stressors–stress–coping–performance associations in construction not only lays a solid platform to further similar studies (fostering the development of stress management research in construction), but also fits into the current knowledge gap of psychological health in occupational safety and health (OSH) education in construction (fostering evolution of OSH education in construction). For the latter, the book benefits individual construction personnel by the overview and analyses of various stress management strategies, and also facilitates the development of stress management interventions in construction. This is predicted to improve the holistic performance and productivity of individual construction personnel, which are key antecedents of project and organisational success.

Our goal is that this book will facilitate the development of stress management research and education in construction, while also enhancing the awareness of construction personnel on the importance of stress and stress management.

Mei-yung Leung,
Isabelle Yee Shan Chan
and Cary L. Cooper

Acknowledgements

This book involved many people beyond the authors. The authors would like to express special thanks to colleagues and friends who assisted and contributed greatly in the book writing process: Mr Qi Liang and Mr Sherwood Yu – City University of Hong Kong, Dr Wilson Wan – New City Construction Company Limited, Hong Kong; and Dr Jingyu Yu – Hefei University of Technology, PRC. Their generosity is greatly appreciated.

We also thank Ms Madeleine Metcalfe of Wiley Blackwell for her contributions and patience in all phases of this project. Mrs Umamaheswari Chelladurai, Ms Audrey Koh and Ms Harriet Konishi, also from Wiley Blackwell, and Mrs Gerry Wood at Lancaster University greatly expedited the final stages of putting the book into print.

Construction Personnel in Practice

This first chapter introduces the background and objectives of the book. It also discusses the characteristics of the construction industry; the nature, functions and interrelationships of various construction organisations, projects and personnel from an international perspective; the contribution of various member throughout the construction life cycle; and the potential for these industrial, organisational, project and individual characteristics, as well as construction-related work tasks, to cause stress for construction personnel. Along with the discussion of stress management in construction in the chapters which follow, several studies of construction personnel from various nations will be discussed here in order to present an overview of the current environment, results and trends in stress management research in the construction industry context. This chapter provides background information to contextualise the discussion in the remainder of the book.

1.1   Background to Stress Management in Construction

The construction industry can be characterised as *competitive, dynamic and challenging*. A construction project is a unique human endeavour which combines the different goals and objectives of multiple stakeholders. They need to deploy various resources to tackle change and uncertainties and complete the work within a limited time and specific scope (Turner 1993). Normally, construction personnel working in different organisations (such as clients, government departments, consultancy companies, contractors, subcontractors and suppliers) need to work together to ensure the success of a project. Communication and cooperation between stakeholders is critical for

Stress Management in the Construction Industry, First Edition.
Mei-yung Leung, Isabelle Yee Shan Chan and Cary L. Cooper.
© 2015 John Wiley & Sons, Ltd. Published 2015 by John Wiley & Sons, Ltd.

construction personnel, since this is directly related to their efficiency and the success of the project. However, due to the fragmented and dynamic nature of the construction sector (The Chartered Institute of Building [CIOB] 2002), communication between stakeholders is often limited, causing difficulties in cooperation and resulting in a negative impact on outcomes (Hewage, Gannoruwa and Ruwanpura 2011). Moreover, construction personnel often work in crisis-ridden site environments at high risk of injury. Hence, it is not surprising that the majority of personnel, including architects, project managers, engineers, surveyors and construction workers, report feelings of *stress* in their daily working lives (Leung, Chan and Yuen 2010). According to a recent study of occupational stress in the construction industry (CIOB 2006), almost 70% of personnel suffer from stress, anxiety or depression.

The level of stress experienced by construction personnel largely depends on the source of stress, or more specifically the *stressors*. Stress occurs when an individual encounters a misfit between his/her actual ability and the environment (such as work and home) (French, Rogers and Cobb 1974), which may induce both emotional and physical symptoms (Moorhead and Griffin 1995). Common stressors found in the construction industry include long working hours, tight project schedules, limited organisational support and safety issues (Goldenhar, Williams and Swanson 2003; Leung et al. 2005a). These stressors and the stress reactions they create, are likely to induce physical and mental fatigue, undermine team performance, reduce working abilities and ultimately lead to the potential breakdown of the whole project (Ng, Skitmore and Leung 1995; Sutherland and Davidson 1989).

Stress can have serious implications not only for health and task performance, but also profitability and organisational development (Cooper and Dewe 2008). In terms of health, stress is known to be related to heart disease, negative emotions and extreme physiological symptoms (Cooper 2001). The British Heart Foundation suggests that coronary heart disease, which is often attributable to stress, costs £200 per employee per year in the UK (Hibbert 2003). According to the Health and Safety Executive (2007), 10.5 million working days are lost due to stress-related illnesses in the UK every year. The cost implication of occupational stress for employers is $381 million per annum (Cousins et al. 2004). Moreover, individuals in stressful situations tend to make rigid, simplistic and superficial decisions (Cherrington 1994), leading to the failure of construction projects to meet time, quality and budget requirements. In fact, highly stressed workers change jobs more frequently than those who are under less pressure, so excessive stress may also have a serious effect on the turnover intention of construction personnel (Djebarni 1996).

However, it seems that most construction employers focus mainly on profit while ignoring the importance of the health of their employees. This has a direct influence on project and organisational outcomes. Employees are the most valuable asset of all construction companies (Rowley and Jackson 2011). Ignorance of the effects of stress on their health not only costs companies a lot of money, but can also lead to lawsuits. For example, French, a UK police officer and his colleagues sued their employer, the Chief Constable, for negligently failing to provide proper training resulting in psychiatric injury (e.g. All England Reporter 2006). Employers have a legal responsibility to provide a healthy and safe working environment for their employees.

However, stress is not necessarily always harmful. When talking about stress, people generally mean "overstress" (too much) and its effects. Though overstress can result in burnout, too little stress (understress) can also affect the performance of construction personnel through the phenomenon termed "rustout" (Lingard 2003). Studies suggest that there is an inverted U-shaped relationship between the degree of stress and level of performance (Leung et al. 2005b; Yerkes and Dodson 1908). To optimise performance, an individual needs to be able to react and cope with his/her stress. There are various coping approaches, from which three main sets of behaviours have been identified: (i) problem-based coping, which focuses on task situations or problem solving (Carver, Scheier and Wein 1989; Weatherley and Irit 1996); (ii) emotion-based coping, which deals with emotional or anxiety reactions (Haynes and Love 2004; Lazarus and Folkman 1984); and (iii) meaning-based coping, which focuses on the attitudes resulting in positive emotions and reenacting the adoption of problem- or emotion-focused coping (Kabat-Zinn et al. 1992). However, not all types of coping strategy are adaptive. Studies in this area tend to categorise problem- and meaning-based coping as adaptive (see e.g. Folkman 2010; Leung, Liu and Wong 2006) and emotion-based coping as maladaptive (Dyson and Renk 2006). Maladaptive coping is not uncommon among construction personnel (Haynes and Love 2004; Leung, Liu and Wong 2006).

In fact, numerous researchers have investigated the stress experienced by different types of professionals such as nurses (Dailey, Ickinger and Coo 1986), teachers (Baruch-Feldman et al. 2002) and general managers (Davidson and Cooper 1983). Stress is also increasingly considered as a major problem for the construction industry in many countries such as Australia, South Africa, the UK and the US (Bowen, Edwards and Lingard 2013; Djebarni 1996; Lingard 2004; Loosemore and Waters 2004). However, little comprehensive investigation into stress management methods for construction personnel has so far been conducted, leading to an information gap. There is clearly a need for a comprehensive research study of effective stress management techniques for construction personnel working in industry.

1.2   Construction Organisations

In the modern industrialised and commercialised world, various organisations are involved in the construction industry, including governments, developers, consultants, contractors, subcontractors, and suppliers.

1.2.1   Governments

In order to keep the construction sector in good order, governments in all countries exert a great deal of influence over various activities. There are generally two types of government structure, namely centralised and decentralised. Each affects the construction industry in their country in different ways.

Centralised government refers to a situation where the government makes all the important decisions affecting the nation or state and handles the associated responsibilities. This type of government can directly or indirectly influence the construction industry by applying a large amount of regulations and policies in the areas of planning, financing, construction and maintenance, with a powerful and immediate effect (Wells 1986). The government can directly enact the industrial regulations and codes which construction personnel must observe. Any violation of the code will attract severe punishment in terms of fines and removal of qualifications for example. In addition, centralised governments can also indirectly affect the construction industry via various policies. For instance, China has a centralised government with a comparatively strong top-down mandate. The policy of opening up the construction market internationally, embarked on in 1992, has attracted various overseas competitors who have introduced new technologies to the Chinese construction market. Such a situation may be stressful for local personnel, who were educated in and have worked under the closed system for years, if they are expected to catch up with new technology and create competitive advantage in the new market environment. Hence, government policy can induce stress in construction personnel (Leung, Sham and Chan 2007).

A *decentralised government* spreads its responsibilities out to different bodies like provinces and states, so they are free to set their own construction laws and regulations and prevent local development from being circumscribed (Wallis and Oates 1998). For instance, in a decentralised national government, it is the local municipal or state government, not the central authority, that is responsible for setting and enforcing the safety, health and environmental standards for building and construction projects undertaken within a city or state. This may place a strain on developers and contractors, especially nonlocal organisations, by for example imposing compulsory regulations and site inspections. It is not easy for construction personnel, particularly those involved in international projects and/or working as expatriates, to understand the specific context of the industry in different states. Unfamiliarity with the building ordinances and regulations in different regions can also cause difficulties at work, once more leading to stress for construction personnel (see e.g. Leung and Chan 2012).

1.2.2   Developers

A developer is usually considered as the party who owns the land, develops it through construction and owns, operates and/or sells the property at the end of the project (Ashworth 2002). Developers hold an important position in the built environment, as they have the responsibility of delivering the enormous amount of real estate required for the growing urban population (Royal Institution of Chartered Surveyors [RICS] 2011). There are three types of developers, classified according to organisational size and capacity: large, medium and small (Anikeeff and Sriram 2008; Psilander 2007). Generally, small and medium developers have fewer resources and capacity to manage risk than the larger ones, which are capable enough to maintain every project in line with governmental regulations. Hence, the former types are more easily affected by changes in government policy or dynamic market situations, such as an increase in inflation leading to project overbudget. Large developers may have sufficient financial and professional support, either inside or outside organisational boundaries. They normally carry out large-scale, complicated projects which involve higher levels of uncertainty and risks from factors such as the economic crisis, government policy and site constraints. Even small changes may significantly affect their profit and overall business strategy. Therefore, different sizes of developers inevitably face different types and levels of stress.

Developers can also be classified into private and (semi-)public (Barrie and Paulson 1992). *Private developers* work on construction projects for private individuals or groups, while *(semi-)public developers* are government-related organisations developing such as hospitals, colleges, schools, libraries and churches, etc. Normally, private developers aim for profit. A developer has to control the project finances closely in terms of various aspects including land cost, design, construction and marketing, in order to ensure business survival. All projects must follow the regulations and obligations created by government. However, government economic policy, housing problems and environmental issues can easily affect operational strategy and profits, thereby inducing stress to project members. Meanwhile, although (semi-)public developers are less susceptible to stress from economic and political changes, they are also responsible for providing appropriate and adequate facilities to satisfy the diverse needs of the public. Such demands are especially challenging in democratic countries, where citizens have more freedom of speech and are more likely to raise complaints and criticisms towards the government.

1.2.3   Consultants

The term consultant usually refers to a professional or a firm in a specific field with a specific knowledge of a particular subject matter and influence over an individual, group or organisation (Block 1999). Various consultants work in the design and management of construction projects, such as planners, architects, structural engineers, civil engineers, building services engineers, quantity surveyors, financial advisors, lawyers, project managers, "green" building consultants and landscape engineers. They tend to have the best-in-class expertise to advise developers and contractors on a diverse range of built environment issues (RICS 2011). Each consultant offers particular management services to ensure that projects can be delivered successfully (Kirmani and Baum 1992). Although consultants normally have less direct power to make major changes to projects in terms of design, programming and cost (New York State Department of Transportation [NYSDOT] 2003; Tordoir 1995; Waters 2004), failure to select appropriate personnel can lead to unsatisfactory project performance during the design, construction, completion or operations stages (Nitithamyong and Tan 2007).

Consultants play essential roles in the different stages of a construction project, from inception to post-construction. In practice, consultants (such as architects and civil engineers) can offer feasibility research services in the predesign stage; prepare drawings, specifications and cost plans in the design phase; advise on selecting the best contractor in the tendering stage; and manage and control quality, cost and programming in the construction phase (Chow and Ng 2010; Smith and Jaggar 2007; Stein and Hiss 2003). In addition, conflict is inevitable in the construction process, creating stress for consultants such as quantity surveyors (Leung, Skitmore and Chan 2007) given that they tend to play an essential gatekeeping role in protecting developers or contractors from the various claims made by other parties. In the post-construction stage, consultants need to carry out inspections in order to ensure the quality of the final products, because they have a professional responsibility to ensure the building satisfies the requirements of the project owner and complies with regulations (RICS 2011). In other words, the work scope of consultants covers every stage of the construction process. However, since they tend to be located low in the project hierarchy with limited direct power, they often face excessive or conflicting demands which increase their workload and extend their working hours, resulting in stress (Leung et al., 2005a, 2008).

1.2.4   Contractors

Builders are usually regarded as *contractors* who deliver a construction product for a specified cost at a particular time (Hans 1984). In general, contractors can be divided into two types, namely main and subcontractors. *Main contractors* participate in a process of transforming inputs (labour, materials, equipment and capital) into outputs (construction products) (Winch 2010). At the preconstruction stage, they will have to compete against each other to win the contract. Price and other factors, including experience, reputation (such as safety record) and professional qualifications, are considered critical selection criteria. Main contractors often put themselves in fierce competition so as not to lose out on market opportunities. During the construction process, main contractors need to carefully plan and manage all the necessary resources for erecting a building (such as the construction phases, sequences, programming and budgeting), which involves an immense amount of work. Inappropriate management can lead to the failure of construction projects as a result of problems such as exceeding cost limits, poor-quality buildings, safety incidents and delays (Clough and Sears 1991). Moreover, there are lots of stakeholders involved in a construction project, such as the client, planners, project managers, architects, engineers, surveyors, suppliers, specialists and end users (Chinyio and Akintoye 2008). Main contractors have to liaise with, coordinate and manage all these parties so as to ensure successful delivery. Hence, given this heavy responsibility and workload, the staff of main contractor firms can easily suffer from stress (Djebarni 1996; Leung, Chan and Dongyu 2011).

A *subcontractor* denotes a construction firm that is responsible for undertaking part of the work on behalf of the main contractor (Errasti et al. 2007). In general, there are two types of subcontractor in the construction industry: (i) domestic subcontractors, who are selected and employed directly by main contractors in a private arrangement and have no contractual relationship with the client; and (ii) nominated subcontractors, who tender to and are then appointed and employed by the architect or client. In the latter situation, the main contractor will be instructed by the client to enter into a contract with the nominated subcontractor (RICS 2011). Domestic subcontractors are often considered to be effectively part of the human resources of the main contractor. They have comparatively less protection from the client and cannot enjoy the benefit of the fluctuation clause in the contract between the main contractor and client, inducing financial pressure and the stress associated with this. On the other hand, nominated subcontractors also face difficulties such as disputes over the release of retention monies and mutual interference with the main contractor in the construction process. In addition, main contractors may realise the greatest cost savings by withholding from subcontractors, increasing the prevalence of unfair practices (Kennedy, Morrison and Milne 1997). Hence, nominated subcontractors are also in a stressful situation.

1.2.5   Suppliers

For the delivery of construction products, a supply chain should be established in which resources, including labour, materials and equipment, are organised and utilised (Winch 2001). Suppliers can be classified into two main groups: materials and equipment. Both play a major role in a construction project, as the condition and quality of materials and equipment significantly affect success. *Materials suppliers* usually supply the raw materials for the delivery of construction projects and also assist the main contractor in inspecting quantity and quality. On the other hand, equipment, such as compactors, excavators and tower cranes, is also necessary for construction projects. This is provided by *equipment suppliers* (Day and Benjamin 1991). To ensure the requirements and contractual specifications are met, equipment suppliers usually provide onsite installation and regular inspections and sometimes also carry out onsite maintenance.

To achieve project outcomes successfully, it is necessary to select proper suppliers of both materials and equipment. In the public sector, completeness and fairness are considered to be important. Hence, all potential suppliers are identified and given equal opportunity to tender for the work. Various rules, norms, standards and legal requirements have to be reviewed, while a systematic assessment approach, such as a weighting scheme, is applied to regulate the process and enhance transparency, objectivity and nondiscrimination (Telgen and Schotanus 2010). In the private sector, profit maximisation is normally the main goal of developers and contractors (Wallace 2007). Selection of suppliers is therefore conducted only when there is a real need to identify the best. In most situations, suppliers compete in terms of various aspects such as minimising material and production costs while simultaneously enhancing the quality of services and products. The final cost always fluctuates depending on the market (e.g., an increase in material prices may cause the supplier to lose all marginal costs). Furthermore, since most construction projects have a tight schedule, any delay in the delivery of supplies might cause severe problems for the project owner and other personnel. Hence, material and equipment supply is often in the critical path of the project, which makes it unavoidably stressful for those involved.

1.3   Construction Personnel

According to the outline plan of work issued by the Royal Institute of British Architects (2007), construction work involves four major stages: briefing, sketch plans, preparing working drawings, and site operations. The second and third of these are closely related to project design and can be collectively termed the *design stage*, while the site operations phase is also known as the *construction stage*. Construction personnel contribute their particular knowledge to each phase in order to ensure project success.

1.3.1   Project Managers

Project managers, working on behalf of the project owner, are responsible for getting the project completed in accordance with the schedule, budget and quality standards. They achieve this by organising, planning, scheduling and controlling the field work (Sears, Sears and Clough 2008). They act as the central communication point between the project owner, design team, contractors and other parties and engage in the various activities involved in construction projects from inception to completion. In the *briefing stage*, the project manager is responsible for feasibility studies, site visits and the employment of consultants (CIOB 2000). In the *design stage*, he/she needs to advise clients on selecting the procurement method, managing the design process and selecting contractors (Bennett 2003). During the *construction stage*, the project manager acts as an effective leader and is responsible for the implementation of the programme, arrangement of project finances, fulfilment of statutory requirements and so on (CIOB 2010). Moreover, managers also need to ensure that all construction work adheres to environmental protection ordinances and that all onsite staff are working in a safe environment with adequate protection from injury or accident.

As the leaders of construction projects, project managers are regarded as key personnel and their performance significantly affects the whole project. Their roles include not only planning, organising and supervising the project team and overseeing the progress of work, but also dealing with demanding time pressures, intrinsic uncertainties and multiple stakeholders (Leung, Chan and Dongyu 2011). Hence, they normally have an intensive workload over a lengthy period, since they are responsible for the project from the briefing to the construction stages. In addition to this heavy workload, project managers also shoulder lots of responsibility, as every single decision they make has a direct impact on the scheduling, cost, quality and/or ultimate success of the project. Therefore, it is inevitable that they are subject to a great deal of stress. The decision making of individuals working in excessively stressful situations is generally more rigid, simplistic and superficial (Cherrington 1994). Excessive stress has been shown to have a negative impact on the performance of project managers in the construction industry, which in turn has a negative impact on outcomes (Leung, Chan and Olomolaiye 2008).

1.3.2   Architects

Architects are licensed professionals trained in all aspects of the art and science of building design (Bennett 2003; Jackson 2004). They can also act as the controllers of the construction process in order to ensure that the building is built as designed (Morton 2002). In the *briefing stage*, the architect carries out studies of site conditions and project issues and provides a professional opinion about project feasibility to the owner. Architects generally also provide solutions regarding building design. In the *design stage*, architects are mainly responsible for the preparation of detailed drawings and specifications and also participate in the tendering process (Chappell and Willis 2010; Ogunlana 1999). In addition, architects also need to communicate with other consultants in order to solve all architectural, legal, technical and financial problems arising; meet the client's requirements; and prepare a full scheme for statutory approval. They also need to supervise the actual *construction stage* by regularly inspecting the construction work and issuing instructions to ensure that it is being performed effectively. They also take responsibility for resolving disputes among various other parties (Mow 2006).

Playing such a wide range of project roles means architects face a heavy and challenging workload. Not only do they need to meet all the requirements of the project owner (e.g. function, value, cost), but they also need to ensure the design does not violate mandatory government building regulations and laws (such as town planning requirements, fire safety design requirements, height limitations) (Oyedele and Tham 2007). In addition, architects also need to deal with the complicated relationships among various parties in order to keep the focus of the entire project team on achieving its goals. Like project managers, an architectural professional plays a long-term and challenging role in a construction project, from inception to completion. Since building designs can be affected by numerous constraints discovered by various professionals at different stages (such as subsoil constraints detected by geotechnical consultants, difficulties in building services design in the limited space above the suspended ceilings as encountered by building services engineers and over-budgeting as estimated by quantity surveyors) architects are driven by various pressures to liaise and cooperate with different parties. They may also be required to revise their designs in different construction phases, which is a particularly stressful element of their role.

1.3.3   Engineers

Engineers are usually the lead designers for infrastructure and civil projects (Jackson 2004). They design materials, structures and systems in consideration of the project limitations, feasibility, safety and cost. *Civil engineers,* including structural, material and geotechnical and building services engineers (such as fire services, mechanical, lift and electrical engineers), are the major types found in the construction industry (Schwartz 1993). *Civil engineers* are responsible for designing suitable structures to support the load of construction products, such as bridges, tunnel dams and buildings. They usually need to visit the site to understand the natural environment in the *briefing stage,* so that they can design the most suitable structure (Behrens and Hawranek 1991; Ng and Chow 2004). Then, civil engineers will cooperate with other professionals to prepare drawings and specifications in the *design stage* (Lupton 2001). In the *construction stage,* they must assist other parties to resolve any technical problems and ensure the quality of construction products. As well as affecting the shape and layout of the construction project, the work of civil engineers directly determines the stability, reliability and safety of the structures being built, so any mistakes may cause significant losses in terms of money and human lives (Building Department 2010; Manik, Greenhouse and Yardley 2013). Such design work is difficult for civil engineers, especially on uncertain soil conditions, which undoubtedly causes them high levels of stress.

Building services engineers are responsible for the design, installation, operation and monitoring of the mechanical and electrical systems used in buildings (Chappell and Willis 2010; Muscroft 2005; Shoesmith 1996). Their work focuses mainly on the safe, comfortable and user-friendly operation of the final products. In practice, different types of building services engineers will work together in order to provide suitable facilities within a limited space. A well-planned, compact building services design not only enhances the quality of life of building occupiers, but can also leave more space for profit making (Shoesmith 1996). In the *briefing stage,* building services engineers investigate the microclimatic conditions of the proposed conceptual design and prepare preliminary building services design criteria. Based on these criteria and with a consideration of life-cycle costs, they then design the lighting, heating, air-conditioning and other systems in the *design stage.* In addition, they need to ensure the correct installation of the various systems in the *construction stage* and advise on maintenance requirements. They are also responsible for providing services which meet practical requirements and avoiding any violation of mandatory government regulations (such as those relating to fire safety and air ventilation). Any deviation may induce additional costs and/or time. In fact, since building services normally account for over 25% of total construction costs (Census and Statistics Department 2011), they are critical to the overall construction programme. Holding such a critical position, while at the same time being caught in the middle of the conflicting expectations of clients and architects and being subject to the requirements specified by building ordinances, building engineers can also suffer from significant stress.

1.3.4   Quantity Surveyors

Quantity surveyors are responsible for studying the economic and financial implications of a construction project and advising the project owner, the architect and engineers on matters relating to cost (Ashworth and Hogg 2002). All of their work aims to ensure that the resources of the construction industry are utilised to the best advantage of society (RICS 1976). The responsibilities of quantity surveyors include a variety of activities, from inception to completion. In the *briefing stage,* quantity surveyors appraise the local construction cost level, collect cost information from similar projects and so on, in order to advise the project owner on appropriate cost limits (Ferr, Brandon and Ferry 1999). Then, with the cooperation of other participants, they prepare a cost plan to guide the design work of architects and engineers, write tendering documentation to use when selecting contractors and undertake various other tasks in the *design stage* (Ferry, Brandon and Ferry 1999). During the *construction* stage, quantity surveyors prepare valuations for interim certificates, submit financial statements for variation and outstanding claims made by contractors and prepare the final accounts (Fryer, Fryer and Egbu 2004).

In practice, quantity surveyors are often pressed to produce accurate cost estimates within a rigid but short timeframe. The accuracy of their budgeting directly affects the outcome of the project and any over- or underestimates can be costly for the owner. This difficult task is often made even harder by a lack of cooperation between various project team members, such as project managers, main and subcontractors and suppliers, given that cost estimation relies heavily on the quality of the data associated with the production schedule, methods, materials, quantities and component costs. In such circumstances, it is not surprising to find that quantity surveyors are often under a considerable amount of stress (Leung, Chan and Olomolaiye 2008). On the other hand, their stress levels may escalate towards the end of a project, since they are responsible for signing off diverse variation orders and claims throughout the process from the beginning to the maintenance period.

1.3.5   Construction Workers

Construction workers are skilled and unskilled labourers who undertake a range of manual jobs. They are the front-line force of construction projects (Anwar 2004; Fung et al. 2005). Rather than being a single class of skilled and factory workers, construction workers engage in various trades such as excavation, concreting, brickworks, plastering, carpentry and pumping (Vocational Training Council and Building and Civil Engineering Training Board 2011). They have the most significant and direct impact on project outcomes in terms of time, cost, quality, safety and environmental impact (Applebaum 1999). Compared to other participants, construction workers mainly contribute physical work onsite. Due to the poor working environment and the physically demanding nature of their efforts, construction is a very high-risk environment for the individual worker. Construction workers run a higher risk of injury than those in other industries through accidents such as falling from height, being hit by falling objects from above, being injured by equipment such as hoists and steel-cutting machines and suffering electric shocks. In fact, construction has long been recognised as one of the most dangerous occupations (Beswick et al. 2007), in terms of both its annual industrial accident (63%) and fatality (79%) rates. These are the highest of all industries in various cities and countries like the US (Mitropoulos and Memarian 2012), UK (Zhou, Fang and Wang 2008) and Hong Kong (Census and Statistics Department 2009).

In addition to their exposure to the risk of accidents, construction workers also suffer from high levels of stress. Their work frequently requires prolonged standing, bending, working, lifting and carrying heavy things in a confined area either inside or outside the structure. Moreover, construction workers are often positioned at the lowest level of an organisation and have limited power over their resources and goals. It can be very difficult for them to reflect their needs and negotiate their rewards at the organisational level (see e.g. Tabassi and Bakar 2009). All these factors can cause stress. It has been shown that construction workers have the third-highest stress levels of any occupation worldwide and suffer from 1.7 times more stress than personnel in other industries (International Labour Organization 1992). As a consequence, numerous working days and millions of dollars are lost every year due to the negative consequence of stress among construction workers (Health and Safety Executive 2007).

1.4   Construction Projects

1.4.1   Nature

A project is a series of interrelated activities undertaken to create a unique product, service or result within a specific timeframe and under certain constraints (PMI 2000). A construction project usually also has particular attributes such as a long life cycle, capital investment, a large number of activities and participants, close regulation by government and the need to manage the effect of the environment and weather on operations (CIOB 2010). There are many different kinds of construction projects (Bosch and Philips 2003), which can be grouped into two broad categories: building and infrastructure. A *building project* is a way to add small or large structures to real property and land, including residential, commercial, industrial and other types (Bennett 2003). Such projects normally involve the construction of many different items such as external and internal walls, floor, wall and roof finishes, doors, windows, ironmongery and sanitary fittings. The cost of building projects usually varies significantly based on various factors such as the market situation, project locations, materials and the requirements of the project owner (Clough and Sears 1991; Martin 2004). Most are undertaken for the purpose of profit, so cost control may be overemphasised. The plethora of techniques which may be applied to control cost and maximise profits can increase the workload of various participants and place them under stress.

Infrastructure projects are those which build up the infrastructure of a country, including highway, tunnels, bridges, dams, restraining walls, drainage, slop surfaces, airports, harbours, waterways, dikes and retention ponds and pipelines (Bosch and Philips 2003; Plunkett 2007). Their development often requires multiple factors to be considered in terms of their design, financial, legal and social aspects. Normally, they are not undertaken for profit, but in the public interest. Budgets must be approved via a series of formal governmental steps. Mega infrastructure buildings demand sophisticated building techniques (such as geotechnical works for uncertain tunnels and slop projects, aerial technology for terminal buildings, runways and aircraft parking areas in airport projects). Furthermore, some construction projects (e.g. those involving highways, tunnels, airports and water supplies) are located in areas such as mountains and deserts which are far away from towns and cities, leading to a particularly tough working environment for those involved.

1.4.2   Construction Procurement

Procurement is the process used to deliver construction projects (Ashworth and Hogg 2002). Project procurement in the construction industry involves a systematic process of identifying and obtaining, through purchase or acquisition, the necessary project services, goods, or results from the outside vendors who will carry out the work. The selection of appropriate procurement methods is therefore essential as it determines the framework for project management, as well as the timetable, quality control and cost plans. All of those elements directly influence the duties of construction personnel and the risks they face and exert a degree of pressure and stress in the work environment.

Various methods of procurement are used in the construction industry, such as traditional, design and build, construction management, management contracting, build–operate–transfer (BOT) and build–own–operate–transfer (BOOT) (Cook and Williams 2004). Each method has advantages and disadvantages for the different parties involved. There is no single approach which always outperforms others in all situations, because each method brings a different degree of uncertainty and risk to each project. Adopting traditional procurement methods, for instance, involves various personnel in a specific sequence in which the project manager, architect, engineers and quantity surveyor will be employed to produce tender documents for use in selecting the contractor. Although this method affords the project owner a high level of cost certainty for a relatively low tender price (Ashworth and Hogg 2002), it may impose time pressures and hinder synergy arising from cooperation between different parties at the design stage, stifling innovation. In fact, lack of participation by the contractor in the design process may lead to low constructability of design, leading to further problems such as scope variations, conflict and disputes of various types. This is a further potential source of stress for those involved.

As an alternative to the traditional procurement method, *design and build*, which has now become one of the most popular approaches, is widely used in the construction industry (Quatman 2000). This arrangement transfers maximum risk to the contractor and shortens the overall project timescale, as the contractor takes full responsibility for the project involving both design and construction tasks. In line with having less responsibility and risk, however, the project owner may also have less opportunity to participate in the design stage and less power to control the design. This may lead to conflict and reduce the quality of outputs. In order to monitor project performance, project owners often hire an architect and quantity surveyor to provide essential independent advice and assess contractors' proposals. This directly affects progress and even overall cost as priced by the contractor. All parties may come under pressure from their own organisations, inevitably leading to stress.

1.5   Stress in the Construction Industry

Construction projects normally involve multiple stakeholders, long time-scales and substantial investment. Delivery consists of various complicated tasks. Every project is unique and dynamic and there are always unexpected problems (Langan-Fox and Cooper 2011). Construction personnel need to work with different team members across various projects at different stages. They can easily face **stress** during the implementation period and experience physical and mental stress symptoms such as depression, anxiety and poor physical health (Leung, Skitmore and Chan, 2007). Stress is prevalent in the construction industry. In a CIOB survey (2006), almost 70% of respondents admitted that they suffered from stress and more than half felt the construction industry had become more stressful. Moreover, the findings of existing studies may not fully reflect the real situation. Due to the impact of a popular 'macho' culture, construction personnel may not perceive stress as a problem (Gunning and Keaveney 1998) or may intentionally conceal their stress levels because they do not want to be regarded as weak (Health and Safety Executive 2007).

Studies show that cost, time and other people are the primary sources of stress for construction personnel (Sommerville and Langford 1994). For instance, dealing with changes to projects requires high-quality team performance (Loosemore and Waters 2004) and results in increased pressure to complete the project on time. Long working hours also cause high levels of stress (Lingard 2004) and the poor physical working environment in the construction industry also affects all participants and causes safety problems (Goldenhar, Williams and Swanson 2003). Moreover, being forced to finish complicated projects on time to the required quality and cost standards, as well as meeting the other subjective and objective requirements of the job, can be another source of stress (Haynes and Love 2004). In order to complete their work, construction personnel are assembled in teams in which they must cooperate and where their actions affect one another. Developing trust, collaboration, respect and conflict management are the responsibility of all team members (Adams and Galanes 2003) and the need to address these issues has been linked to work-related stress and the psychological and physical health of construction personnel (Cotton and Hart 2003).

More and more attention has been paid over the last 10 years to studying work-related stress in the worldwide construction industry, with studies having been conducted in South Africa (Bowen, Edwards and Lingard 2013), the UK (Goldenhar, Williams and Swanson 2003), Australia (Haynes and Love 2004), the Netherlands (Janssen, Bakker and de Jong 2001), Hong Kong (Leung, Wong and Oloke, 2003; Leung, Liu and Wong 2006; Leung, Chan and Yu 2012) and the US (Loosemore and Waters 2004). Summarising the results of these studies (see Table 1.1) not only helps us to begin developing an integrated stress management model (Stressors–Stress–Coping Behaviour–Performance) which will benefit both construction personnel and academics working in related fields, but also serves as a platform for identifying and directing future trends in stress management research in the construction sector.

To develop a conceptual model for stress management in the construction industry, a systematic review has been conducted to summarise the outcomes of the various relevant research studies based on a coherent plan and search strategy. Studies were selected by identifying keywords such as 'stressors', 'stress', 'coping', 'performance/outcome', and/or 'construction' in the title, abstract and/or keywords of papers. To ensure quality, only papers listed in PsychINFO, Scopus and the Social Science Citation Index were included. A total of 45 papers investigating stress among different construction personnel such as project managers, architects, engineers, quantity surveyors and construction workers worldwide were included in the review. As shown in Table 1.1, this includes seven papers focused on the targets of project managers (one in South Africa, one in Australia, two in Hong Kong and three in the UK); nine studies investigating construction workers (one in Spain, one in Australia, two in the Netherlands, two in the UK and three in Hong Kong); and 29 papers investigating construction professionals in general (one in the US, two in the UK, three in South Africa, five in Australia and 18 in Hong Kong). Moreover, a further survey conducted by the CIOB covered a broad range of construction personnel from onsite workers to professionals.

Studies to date have tended to describe the degree and types of stress experienced by construction personnel working in industry. Construction personnel often suffer from anxiety and depression (Haynes and Love 2004; Sutherland and Davidson 1993) or even burnout (Janssen, Bakker and de Jong 2001; Leung, Skitmore and Chan 2007; Lingard 2004; Yip and Rowlinson 2009a, 2009b). Apart from the mental and psychological strain involved, construction personnel can also suffer from physical problems (Goldenhar et al. 1998; Goldenhar, Williams and Swanson 2003; Leung, Chan and Olomolaiye 2008). Out of the 45 papers studied here, nine classify the different dimensions of stress, including objective and job-related, subjective and emotional and physical (Leung et al. 2008; Leung, Chan and Yu 2009; Leung, Chan and Yu 2012).

Table 1.1 Systematic literature review of stress management studies in the construction industry setting.

Source	Location	Sample	Variables	Findings
Sutherland and Davidson 1989	England	36 male middle and senior construction site managers	Stressors (communication problems, work overload, conflict, work–home conflict) Mental health Anxiety Job satisfaction	1. Job satisfaction levels of construction site managers were not as high as for comparable managerial groups and most dissatisfaction was related to employee relations issues 2. Mental well-being for construction site managers was lower than for other population groups 3. Social support from a spouse or partner → anxiety (–), depression (–) and mental well-being (+)
Sutherland and Davidson 1993	England	Questionnaire: 561 construction site managers; Interview: 36 male middle and senior construction site managers	Task, personal, organisational stressors Mental health Anxiety Job satisfaction	1. Job satisfaction levels were low compared to a normative population and influenced by grade level 2. Mental health was affected by both grade of management and type of contract 3. Anxiety levels were significantly high, independent of managerial grade 4. Overload and role insecurity (fear of failure) → mental health (–) and high anxiety (+) 5. Organisational culture and climate → job dissatisfaction (+).
Djebarni 1996	Algeria	71 site managers	Stress (boss, job and environment–job) Performance (leadership and project effectiveness)	1. Job stress → performance (inverted U-shape). 2. Boss stress → performance (–) 3. Environment-job stress → performance (–)
Goldenhar et al. 1998	England	211 female construction workers	Job/task demands (job demands responsibility for safety of others, overcompensation, control) Organisational factors (harassment and discrimination, job certainty, availability of training, safety climate, skill utilisation)	1. Responsibility for others' safety and support from supervisors and male coworkers → job satisfaction (+) 2. Increased responsibility, skill utilization, harassment and gender-based discrimination and overcompensation → psychological symptoms of stress (+) 3. Overcompensation and job certainty →insomnia (+) 4. Harassment and discrimination → nausea and headaches (+)

Author/year	Country	Sample	Variables	Findings
Janssen, Bakker and de Jong 2001	Netherlands	210 construction workers	Physical conditions (physical and chemical exposure) Moderator (social support from supervisor and male coworkers) Psychological symptoms of stress (affect: tense, sad, angry) Physiological symptoms of stress (headache, insomnia, nausea) Job satisfaction Illness (hypertension) Injuries (preventable events and near misses) Job demand, job control, social support Burnout	1. Lack of social support → burnout and health complaints among construction workers (+) 2. A significant three-way interaction effect partly confirmed the synergism hypothesis: physical demands were only related to burnout if participants had poor job control and reported high social support
Goldenhar, Williams and Swanson 2003	England	408 construction labourers (male and female)	Job stressors (job demands, job control, job certainty, overcompensating at work, safety climate, skill training, skill underutilisation, responsibility for the safety of others, safety compliance, exposure hours, harassment and discrimination and social support) Physical symptoms Psychological strain Injury Near-miss outcomes	1. Job demands, training, responsibility for the safety of others and safety compliance → injury, near misses (−) 2. Job control, job certainty, safety climate, skill underutilisation, exposure hours → injury, near misses (+) 3. Harassment /discrimination, job demand, job certainty, safety compliance → physical symptoms (+); compliance → physical symptoms→ injury accidents (+) 4. Harassment /discrimination, job certainty, social support, skill underutilisation, safety responsibility, safety compliance→ psychological strain → near misses (+)

(Continued)

Table 1.1 (Continued)

Source	Location	Sample	Variables	Findings
Lingard 2003	Australia	182 engineers	Job stressors (overload, responsibility, role clarity, role conflict, satisfaction with promotion, social satisfaction, job control and satisfaction with pay) Personality traits (neuroticism, extroversion, optimism, quick-wittedness and impulsiveness) Burnout (emotional exhaustion, cynicism, personal competence and professional worth) Turnover	1. Overload, role conflict, neuroticism, quick-wittedness, role clarity, promotional satisfaction → burnout (emotional exhaustion, professional worth) (+) 2. Promotional satisfaction, responsibility, extroversion, role clarity, pay satisfaction, neuroticism → burnout (cynicism, personal competence) (−) 3. Cynicism, emotional exhaustion → turnover intention (+)
Leung, Wong and Oloke 2003	Hong Kong	90 cost estimators	Personal background information Stress Problem-focused coping behaviours (control action, instrumental support seeking, preparatory action) Emotion-focused coping behaviours (escape, emotional discharge, religious emotional support)	1. Senior estimators normally apply problem-focused coping, while junior estimators prefer to seek emotional support from their religion 2. Senior estimators: instrumental support seeking → stress (−) 3. Junior estimators: religious emotional support → stress (−)

Haynes and Love 2004	Australia	100 project managers	Coping strategies Affect state Stress, depression, anxiety	1. Work experience (−), project size (−), avoidance (+), active coping (−), positive affect (−), negative affect (+) → depression 2. Age (+), work experience (−), project value (−), social coping (+), avoidance (+), active coping (−), accepting (+), negative affect (+) → anxiety 3. Education level (+), work experience (−), project size (−), social coping (+), active coping (−), self-control (+), negative affect (+) → stress 4. Respondents who engaged in a more problem-focused style of coping were better adjusted than those who engaged in more emotion-focused behaviours
Lingard 2004	Australia	182 civil engineers	Job stressors Relationship satisfaction Burnout	1. Relationship conflict (+), overload (+), role conflict (+), promotion satisfaction (−), role clarity (−), responsibility (−), job security (+) →burnout 2. Burnout was predicted by different variables among respondents in dual compared to single-income households and among parents and nonparents
Loosemore and Waters 2004	US	84 male and 47 female construction professionals	Factors intrinsic to job Managerial role factors Relationship to others Career and achievement factors Organisational structure and climate	1. Men experience slightly higher levels of stress than women 2. Men appear to suffer more stress in relation to risk taking, disciplinary matters, implications of mistakes, redundancy and career progression 3. Female professionals often suffer from a lack of opportunities for personal development, poor rates of pay, keeping up with new ideas, business travel and the cumulative effect of minor tasks 4. These differences reflect women's traditional and continued subjugation in the construction industry
Siu, Philips and Leung 2004	Hong Kong	374 construction workers	Safety climate (safety attitudes and communication) Psychological strain (psychological distress and job satisfaction) Safety performance (accident rates and occupational injuries)	1. Safety attitudes → occupational injuries (−) 2. Psychological distress → occupational injuries (+) 3. Job satisfaction → occupational injuries (−) 4. Psychological distress was found to be a mediator of the relationship between safety attitudes and accident rates

(Continued)

Table 1.1 (Continued)

Source	Location	Sample	Variables	Findings
Ng, Skitmore and Leung 2005	Hong Kong	97 construction professionals	Stressors (related to work nature, work time, organisational policy, organisational position, situational / environmental and relationships)	1. Stressors which are most difficult to manage include bureaucracy, lack of opportunity, work–family conflicts and having different views from superiors 2. The patterns of stress manageability differ between clients, consultants and contractors
Lingard and Francis 2005	Australia	231 male construction professionals	Job demand (work hours, irregularity) Work–family conflict (emotional exhaustion, cynicism, professional efficacy) Burnout	1. Work hours, irregularity→ emotional exhaustion dimension (+) 2. Work–family conflict → emotional exhaustion (+) 3. Work–family conflict is a key mediating mechanism between job demands and emotional exhaustion for male employees
Leung et al. 2005a	Hong Kong	180 cost estimators	Stressors (personal, interpersonal, task, physical) Stress	1. Work overload (+), role conflict (+), job ambiguity (+) and working environment (+) → stress 2. Work underload (–) → role conflict (+) → stress
Leung et al. 2005b	Hong Kong	177 cost estimators	Stress Performance (weak interpersonal relationships, organisational relationships, professional performance, ineffective processes and organisational relationship)	1. Stress → weak interpersonal relationships, negative organisational relationship and ineffective process (+) 2. Stress → overall performance (–) 3. Stress → organisational relationship (inverted U-shape)
Love and Edwards 2005	England	100 construction project managers	Job demand Job control Social support (work and nonwork related) Job satisfaction Health	1. JDC-S model can significantly predict employees' psychological well-being in terms of worker health and job satisfaction among the construction project managers sampled 2. Nonwork-related support and work support → health (+) 3. Nonwork-related support and job control → satisfaction (+)
CIOB 2006	England	847 construction participants	Stressors (physical, organisational, job demand and job role)	1. 68.2% of respondents had suffered from stress, anxiety or depression as a direct result of working in the construction industry 2. Lack of privacy (43%), inadequate temperature controls (43%), lack of feedback (56.8%) poor communication (55.7%), inadequate staffing (55%), too much work (64.1%), ambitious deadlines (59.7%), pressure (59.9%) and conflicting demands (52.2%)→ occupational stress

Author/Year	Country	Sample	Variables	Findings
Lingard and Francis 2006	Australia	202 construction professionals	Work–family conflict Support (coworker, supervisory and perceived organisational support) Burnout (emotional exhaustion)	1. Organisational support is a moderator between work–family conflict and emotional exhaustion 2. Supervisory support (practical and criticism) is a moderator between work–family conflict and emotional exhaustion 3. Coworker support (practical) is a moderator between work–family conflict and emotional exhaustion
Leung, Liu and Wong 2006	Hong Kong	210 cost estimators	Coping behaviours Stress Estimation performance (project performance, poor processes, good cooperation and interpersonal relationships)	Senior estimators: 1. Problem-focused coping behaviours (+), emotional-focused coping behaviours (−) → project performance 2. Stress → interpersonal relationships (inverted U-shape) Junior estimators: 3. Problem-focused coping behaviours (+), emotional-focused coping behaviours (+), stress (inverted U-shape) → project performance
Leung, Sham and Chan 2007	Hong Kong	163 cost estimators	Stressors (personal behaviour, role conflict, social support, poor environment, work underload and work–home conflict) Job demand	1. A significant relationship between quantitative and qualitative stress levels 2. Job demand stressors (work underload and work–home conflict) → qualitative job demand stress (+) 3. Qualitative job demand stress → quantitative stress (+) 4. Nonjob demand stressors (personal behaviour, poor environment, social support)→role conflict stressor (+)
Leung, Skitmore and Chan 2007	Hong Kong	163 cost estimators	Stressors (personal, task, organisational and physical) Stress (objective and subjective and emotional exhaustion)	1. The stress levels of both the professional estimators and other personnel are similar, with objective stress being significantly higher than subjective stress, which is in turn significantly higher than emotional exhaustion 2. Lack of autonomy (+), lack of feedback (−) → objective stress 3. Relationship conflict (+), lack of feedback (−), lack of autonomy (+) → subjective stress 4. Lack of autonomy (+), unfair reward and treatment (+) → emotional exhaustion

(Continued)

Table 1.1 (Continued)

Source	Location	Sample	Variables	Findings
Meliá and Becerril 2007	Spanish	105 construction workers	Antecedent variables (leadership, role conflict and mobbing behaviour) Intervening variables (tension and burnout) Outcome variables (perceived quality, psychological health and propensity to leave)	1. Leadership→ burnout (–); leadership→ perceived quality (+) and propensity to leave (–) 2. Role conflict → tension (+); mobbing behaviour → burnout (+) 3. Burnout → perceived quality (–) and propensity to leave (+) 4. Tension → psychological health (–)
Sang, Dainty and Ison 2007	England	Architects	Job satisfaction Physical symptoms of stress Work–life conflict Turnover intention	Female respondents reported significantly lower overall job satisfaction and significantly higher physical symptoms of stress, work–life conflict and turnover intention
Leung, Chan and Olomolaiye 2008	Hong Kong	108 construction project managers	Stress (objective stress, burnout and physiological stress) Performance (task, interpersonal and organisational)	1. Objective stress → interpersonal performance (inverted U-shape), task performance (–), organisational performance (–) 2. Physiological stresses → organisational performance (U-shape relationship) 3. Burnout → task performance (+), organisational performance (U-shape relationship)
Leung, Chan, Chong and Sham 2008	Hong Kong	108 client and 68 contractor cost estimators	Stressors (Type A behaviour, social support, work underload, poor working environment, role conflict and private life) Stress	Stressors → stress 1. Poor environment, role conflict, work underload → stress (+) 2. Social support, Type A behaviour → stress (–) 3. Work underload → role conflict among cost engineers, supervisors and organisation → stress Deviations between two groups of cost engineers: 4. Private life → stress of contractors' cost engineers (+) 5. Closed interactive looping relationship among role conflict, Type A behaviour, work underload for contractors' cost engineers
Leung, Zhand and Skitmore 2008	Hong Kong	73 cost estimators	Stressors (personal, physical, organisational and task) Support (emotional, career development, reward, workgroup and superior) Stress	1. Lack of autonomy (+) and lack of feedback (–) → stress 2. Informal organisational support (emotional, workgroup and superior) → stressors (–)

Author/Year	Location	Sample	Constructs	Findings
Yip, Rowlinson and Siu 2008	Hong Kong	222 construction engineers	Four-factor model of coping strategies: rational problem solving, resigned distancing, seeking support/ventilation and passive wishful thinking Burnout: emotional exhaustion, cynicism and reduced professional efficacy Role overload	1. Problem solving significantly moderated the relationship between role overload and all three dimensions of burnout 2. Moderating effect of resigned distancing and seeking support/ventilation was significant only for emotional exhaustion and cynicism, respectively 3. Passive wishful thinking failed to demonstrate a significant moderating effect on any of the burnout dimensions
Leung, Chan and Yu 2009	Hong Kong	108 construction project managers	Stressors (personal, physical, organisational and task) Subjective and objective stress	Four critical stressors (work overload, poor interpersonal relationships, poor work environment and poor home environment) → subjective and objective stress (+)
Yip and Rowlinson 2009a	Hong Kong	403 construction professionals	Long working hours, role overload, role conflict, role ambiguity, lack of autonomy and job security as significant job determinants of burnout	1. Long working hours, role overload, role conflict, role ambiguity, lack of autonomy and job security → burnout (+) 2. Job redesign contributes to the reduction of burnout
Yip and Rowlinson 2009b	Hong Kong	249 construction engineers	Burnout (exhaustion, cynicism and diminished professional efficacy) Long working hours, qualitative overload, lack of promotion prospects, role conflict, role ambiguity and lack of job security	1. Engineers working within contracting organisations report higher levels of burnout than their compatriots within consulting organisations. The results also showed that burnout could be attributed largely to the stressors associated with job conditions and work environments 2. For engineers in consultant organisations, qualitative overload and lack of promotion prospects → burnout (+) 3. For engineers in contracting organisations, long working hours, role conflict, role ambiguity and lack of job security →burnout (+)
Leung, Chan and Yuen 2010	Hong Kong	142 construction workers	Stressors (personal, interpersonal, physical, organisational and task) Job stress and emotional stress Injuries and accidents	1. Emotional stress → injury incidents (+) 2. Work overload, inter-role conflict, poor physical environment, unfair reward and treatment, inappropriate safety equipment → emotional stress (+) 3. Poor workgroup relationships, work overload, inter-role conflict → job stress (+)

(Continued)

Table 1.1 (Continued)

Source	Location	Sample	Variables	Findings
Leung, Chan, and Chong 2010	Hong Kong	139 construction professionals	Chinese values (interpersonal integration, conservative personality, social conventions, disciplined work ethos) Stressors (task, organisational and interpersonal)	1. Valuing social conventions → role ambiguity, poor workgroup relationships (−) 2. Disciplined work ethos → work overload (+), poor workgroup relationships, work underload (−) 3. Conservative personality → work overload (−) 4. Valuing interpersonal integration → disciplined work ethos (+), social conventions (+) → stressors 5. Poor working environment → poor workgroup relationships, role ambiguity (+)
Love, Edwards and Irani 2010	Australia	449 construction professionals	Stress (self, situational and work) Support (Self, situational and work) Mental health (good and poor)	1. Staff of contracting organisations working onsite reported higher levels of poor mental health and greater work stress than consultants 2. Staff onsite experienced greater levels of self-induced stress, whereas consultants reported higher levels of self- and work support 3. For consultants: work support → poor mental health (−) 4. For consultants: self, situational and work support → good health (+) 5. For contractor staff: self-support → good mental health (+)
Ibem et al. 2011	Nigeria	107 construction professionals	High volume of work, uncomfortable site offices, lack of feedback on previous and ongoing building projects and variations in the scope of work in ongoing projects	High volume of work, uncomfortable site offices, lack of feedback on previous and ongoing building projects and variations in the scope of work in ongoing building projects → stress (+)
Leung, Chan and Dongyu 2011	Hong Kong	108 construction professionals	Stress (job stress, burnout, physiological) Performance (task, interpersonal and organisational)	1. Job stress → burnout (+), task performance (−) 2. Physiological stress → organisational performance (−) 3. Burnout → physiological stress (+), organisational performance (−)
Leung and Chen 2011	Hong Kong	45 cost estimators	Stress (objective and emotional) Commitment (affective, continuance and normative)	1. Stress→ commitment (−) 2. Objective stress → normative commitment (−) 3. Emotional stress (−) → affective commitment (−), normative commitment (−)

Mostert, Peeters and Rost 2011	South Africa	529 construction professionals	Job demand (pressure) Job resources (autonomy, colleague support, supervisor support) Work–home inference (WHI) (positive and negative) Burnout (exhaustion and cynicism) Engagement (vigour, dedication)	1. Job demand → negative WHI (+) and burnout (+) 2. Job resource → positive WHI (+) and engagement (+), while → negative WHI (−) and burnout (−) 3. Negative WHI → burnout (+), while positive WHI → engagement (+)
Chan, Leung and Yu 2012	Hong Kong	40 construction professionals	Stress (physical and emotional) Coping behaviours (problem- and emotion-based) Organisational support (adjustment, career and financial)	1. Hong Kong expatriate construction professionals in mainland China from all groups had experienced stress in their expatriate assignments. 2. Identification of three problem-focused (planful problem solving, confrontive reappraisal and instrumental support seeking) and three emotion-focused (emotional discharge, escapism–avoidance and social support seeking) coping strategies 3. Identification of three forms of organisational support (adjustment, career and financial) 4. Coping strategies and organisational support for expatriate and nonexpatriate groups were compared
Lingard, Francis and Turner 2012	Australia	261 construction workers	Work–family conflict (time- and strain-based work interference with family life and time-/strain-based family interference with work) Control Supervisor support Work–family enrichment	1. Work time demands → time- and strain-based work interference with family life (+), time- and strain-based family interference with work (−) 2. Work–family enrichment → work time control (+), supervisor support (+), time- and strain-based work interference with family life (+), time- and strain-based family interference with work (−) 3. Respondents with high work time demands and low work time control (or low supervisor support) reported the highest levels of time- and strain-based work interference with family life 4. The lowest levels of work interference with family life were reported by respondents whose jobs were classified as low work time demand and high work time control (or high supervisor support)

(Continued)

Table 1.1 (*Continued*)

Source	Location	Sample	Variables	Findings
Leung, Chan and Yu 2012	Hong Kong	395 construction workers	Organisational stressors (unfair reward and treatment, inappropriate safety equipment, provision of training, lack of goal setting and poor physical environment) Stress (emotional and physical) Safety behaviours Injury incidents	1. Safety behaviours (−) and lack of goal setting (+) →injury incidents 2. Emotional stress (inverted U-shape), physical stress (+), inappropriate safety equipment (+)→ safety behaviours 3. Provision of training (+), inappropriate safety equipment (+)→ emotional stress 4. Inappropriate safety equipment (+)→ physical stress
Leung and Chan 2012	Hong Kong	40 construction professionals	Interpersonal, task, organisation and physical stressors	1. Identification of four types of stressors: interpersonal, task, organisation and physical 2. Compares stressors of Hong Kong expatriate construction professionals with that of their compatriots based in Hong Kong 3. A stressors model is developed for Hong Kong expatriate construction professionals
Boschman et al. 2013	Netherlands	262 construction workers and 310 construction supervisors	Psychological work characteristics (job demand, job control, social support, job variety and future perspective) Short-term effects of the work day (need for recovery and fatigue during work) Long-lasting effects (distress, depression and posttraumatic stress disorder (PTSD))	1. Construction workers experienced worse job control, job variety and future perspectives (statistically significant); Supervisors experienced higher job demand and need for recovery (statistically significant) 2. Mental health effects among bricklayers and supervisors, respectively, were as follows: high need for recovery after work (14%; 25%), distress (5%, 7%), depression (18%, 20%) and PTSD (11%, 7%) 3. For both construction workers and supervisors, job demand →depression (+) 4. For supervisors, job control and social support→ depression (+)

Study	Country	Sample	Factors	Findings
Bowen, Edwards and Lingard 2013	South Africa	590 construction professionals	Job demand factors (tight deadlines, long hours, inadequate time, work–family conflict, kept occupied, need to work harder) Job control factors (type and pace of work and environment) Support at work (supervisor and colleague support)	1. Construction professionals in South Africa experience high levels of stress at work 2. Architects suffer from more stress than other professionals (engineers, quantity surveyors and project managers). They experience tight deadlines, long working hours and a feeling of being kept occupied. Architects also have more control over their work types and working environment 3. Female professionals feel more stressed than males
Bowen, Govender and Edwards in press	South Africa	676 construction professionals	Job demand factors Job control factors Support at work Discrimination in terms of salary and job security, based on language, race, religion, gender, sexual preference Organisational climate in terms of interpersonal relationships, compensation for work, job stability, etc. Psychological stress Social stress Physical stress Drug use	1. The terminal consequence of occupational stress is not substance use but rather psychological, physiological and sociological strain effects 2. Organisational climate is largely determined by gender, job demand and control and support factors 3. Age, gender, level of job control and organisational climate are significant predictors of discrimination 4. Psychological strain is significantly predicted by age, job demand and job control factors and by organisational climate 5. Sociological strain is determined by age, job demands, discrimination and psychological strain 6. Age and sociological and psychological strain effects behave as significant predictors of physiological stress effects
Poon et al. 2013	Hong Kong;	32 construction personnel	Bureaucratisation, fatigue, underevaluation of the contribution of safety personnel, rivalry manoeuvring Good working relationships, quality of workforce and company organisational culture	1. Bureaucratisation of the safety management system, underevaluation of the contribution of safety personnel and rivalry manoeuvring have an adverse effect 2. Good working relationships, quality of workforce and company organisational culture → burnout (+) 3. It is worth noting that an individual's ability to dilute and accept the burnout effect plays a key role in their final reaction

On the other hand, 23 of the studies also investigate the sources of stress (*stressors*). The most common stressors studied in the construction industry context are task or job related, including work overload, job demands, job control, tight schedules, compensation imbalance, role and conflict (Bowen, Edwards and Lingard 2013; Goldenhar, Williams and Swanson 2003; Leung et al. 2008; Ng, Skitmore and Leung 2005). Moreover, personal (such as dealing with Type A behaviours, communication problems and work–home conflict), organisational (such as organisational support) and physical stressors (such as a poor working environment) are also addressed (Leung, Chan and Yu 2009; Leung, Chan and Yuen 2010; Lingard, Francis and Turner 2012; Sutherland and Davidson 1993).

To deal with stress in work and daily life, construction personnel need to adopt various *coping behaviours* in the context of project management (Chan, Leung and Yu 2012; Haynes and Love 2004; Leung, Wong and Oloke 2003; Leung, Liu and Wong 2006; Yip and Rowlinson, 2008). Studies show that senior construction personnel under stress tend to apply problem-based coping behaviours (instrumental support seeking, direct/ active action and preparatory action), while junior personnel prefer emotion-based coping (escape, emotional discharge and emotional support) (Leung, Wong and Oloke 2003; Leung, Liu and Wong 2006). These results also indicate that problem-based coping behaviours are more effective in reducing stress and enhancing performance, whereas emotion-based coping actually increases stress (Haynes and Love 2004; Leung, Liu and Wong 2006).

The effect of stress on the *performance* of construction personnel in different disciplines has also been receiving increasing attention of late (Djebarni 1996; Leung et al. 2005b, Leung, Chan and Olomolaiye 2008). Fifteen studies investigate this issue, with four dealing with safety performance in the UK and Hong Kong, three addressing job satisfaction in the UK, two covering turnover intention in Australia and Spain and six papers looking at performance in general in the UK, Australia and Hong Kong. Stress is shown to be negatively related to performance for construction professionals (Goldenhar, Williams and Swanson 2003; Lingard 2003), while an inverted U-shape relationship between stress and ultimate performance has also been identified (Djebarni 1996; Leung et al. 2005b; Leung, Liu and Wong 2006).

References

Adams, K. and Galanes, G.J. (2003) *Communication in Groups*. New York: McGraw-Hill.

All England Reporter (2006) *French and Other vs. Chief Constable of Sussex Police*. Retrieved from http://lexisweb.co.uk/cases/2006/march/french-and-others-v-chief-constable-of-sussex-police, 13 March 2013.

Anwar, C. (2004) *Labour Mobility and the Dynamics of the Construction Industry Labour Market: The Case of Makassar, Indonesia*. Adelaide: Flinders University Press.

Anikeeff, M.A. and Sriram, V. (2008) Construction management strategy and developer performance. *Engineering, Construction and Architectural Management*, 15(6), 504–513.

Applebaum, H.A. (1999) *Construction Worker*. London: Greenwood Press.

Ashworth, A. (2002) *Pre-contract Studies-Development Economics, Tendering and Estimating*. Oxford: Blackwell.

Ashworth, A. and Hogg, K. (2002) *Willis's Practice and Procedure for the Quantity Surveyor*. Oxford: Blackwell.

Barrie, D.S. and Paulson, B.C. (1992) *Professional Construction Management*. New York: McGraw-Hill.

Baruch-Feldman, C., Brondolo, E., Ben-Dayan, D. and Schwartz, J. (2002) Sources of social support and burnout, job satisfaction, and productivity. *Journal of Occupational Health Psychology*, 7(1), 84–93.

Behrens, W. and Hawranek, P.M. (1991) *Manual for the Preparation of Industrial Feasibility Studies*. Vienna: UNIDO.

Beswick, J., Rogers K., Corbett E., Binch S. and Jackson K. (2007) *An Analysis of the Prevalence and Distribution of Stress in the Construction Industry*. Buxton: Health and Safety Executive.

Bennett, F.L. (2003) *The Management of Construction*. Burlington: Butterworth Heinemann.

Block, P. (1999) *Flawless Consulting: A Guide to Getting Your Expertise Used*. San Diego: University Associates.

Bosch, G. and Philips, P. (2003) *Building Chaos. An International Comparison of Deregulation in the Construction Industry*. London: Routledge.

Boschman, J.S., Van Der Molen, H.F., Sluiter, J.K. and Frings-Dresen, M.H. (2013) Psychosocial work environment and mental health among construction workers. *Applied Ergonomics*, 44, 748–755.

Bowen, P., Edwards, P. and Lingard, H. (2013) Workplace stress experienced by construction professionals in South Africa. *Journal of Construction Engineering and Management*, 139(4), 393–403.

Bowen, P., Govender, R. and Edwards, P. (in press) Structural equation modelling of occupational stress in the construction industry. *Journal of Construction Engineering and Management*.

Building Department (2010) *Report on the Collapse of the Building at 45j Ma Tau Wai Road to Kwa Wan, Kowloon*. Retrieved from http://www.bd.gov.hk/english/BuildingCollapseReport_e.pdf

Carver, C.S., Scheier, M.F. and Weintraub, J.K. (1989) Assessing coping strategies: A theoretically based approach. *Journal of Personality and Social Psychology*, 56(2), 267–283.

Census and Statistics Department (2009) *Occupational Safety and Health Statistics Bulletin. Census and Statistics Department*, HKSAR. Retrieved from http://www.labour.gov.hk/eng/osh/pdf/Bull08.pdf

Census and Statistics Department (2011) *Key Statistics on Business Performance and Operating Characteristics of the Building*. Construction and Real Estate Sectors in 2011.Retrievedfromhttp://www.statistics.gov.hk/pub/B10800112011AN11B0100.pdf

Chan, I.Y.S., Leung, M.Y. and Yu, S.S.W. (2012) Managing the stress of Hong Kong expatriate construction professionals in mainland China: Focus group study exploring individual coping strategies and organisational support. *Journal of Construction Engineering and Management*, 138(10), 1150–1160.

Chappell, D. and Willis, A. (2010) *The Architect in Practice*. Oxford: Wiley-Blackwell.

Cherrington, D.J. (1994) *Organizational Behaviors: The Management of Individual and Organizational Performance*. Boston: Allyn & Bacon.

Chinyio, E.A. and Akintoye, A. (2008) Practical approaches for engaging stakeholders: Findings from the UK. *Construction Management and Economics*, 26(6), 591–599.

Chow, L.K. and Ng, S.T. (2010) Delineating the performance standards of engineering consultants at design stage. *Construction Management and Economics*, 28(1), 3–11.

CIOB (2002) *Code of Practice for Project Management for Construction and Development*. London: Chartered Institute of Building.

CIOB (2006) *Occupational Stress in the Construction Industry*. London: Chartered Institute of Building.

CIOB (2010) *Code of Practice for Project Management for Construction and Development*, 4th edn. London: Chartered Institute of Building.

Clough, R.H. and Sears, G.A. (1991) *Construction Project Management*. New York: John Wiley.

Cooke, B. and Williams, P. (2004) *Construction Planning, Programming and Control*. London: Macmillan.

Cooper, C.L. (2001) *Organizational Stress: A Review and Critique of Theory, Research, and Applications*. Thousand Oaks, CA: Sage.

Cooper, C.L. and Dewe, P. (2008) Wellbeing: Absenteeism, presenteeism, costs and challenges. *Occupational Medicine*, 58, 522–524.

Cotton, P. and Hart, P.M. (2003) Occupational well-being and performance: A review of organisational health research. *Australian Psychologist*, 38(2), 118–127.

Cousins, R., MacKay, C.J., Clarke, S.D., Kelly, C., Kelly, P.J. and McCaig, R.H. (2004) Management standards work-related stress in the UK: Practical development. *Work and Stress*, 18(2), 113–136.

Dailey, R.C, Ickinger, W. and Coote, E. (1986) Personality and role variables as predictors of tension discharge rate in three samples. *Human Relations*, 39(11), 991–1003.

Davidson, M. and Cooper, C.L. (1983) *Stress and the Woman Manager*. Oxford: Martin Robertson.

Day, D.A. and Benjamin, N.B.H. (1991) *Construction Equipment Guide*. New York: Wiley.

Djebarni, R. (1996) The impact of stress in site management effectiveness. *Construction Management and Economics*, 14(4), 281–293.

Dyson, R. and Renk, K. (2006) Freshmen adaptation to university life: Depressive symptoms, stress, and coping. *Journal of Clinical Psychology*, 62(10), 1231–1244.

Errasti, A., Beach, R., Oyarbide, A. and Santos, J. (2007) A process for developing partnerships with subcontractors in the construction industry: An empirical study. *International Journal of Project Management*, 25(3), 250–256.

Ferry, D.J., Brandon, P.S. and Ferry, J.D. (1999) *Cost Planning of Buildings*. Oxford: Blackwell.

Folkman, S. (2010) Stress, coping and hope. *Psycho-Oncology*, 19, 901–908.

French, J.R.P., Rogers, W. and Cobb, S. (1974) *Adjustment as a Person–Environment Fit, Coping and Adaptation: Interdisciplinary Perspectives*. New York: Basic.

Fryer, G.B., Fryer, M. and Egbu, C. (2004) *The Practice of Construction Management*. Oxford: Blackwell Science.

Fung, I.W.H., Tam, C.M., Tung, K.C.F. and Man, A.D.K. (2005) Safety cultural divergences among management, supervisory and worker groups in Hong Kong construction industry. *International Journal of Project Management*, 23(7), 504–512.

Goldenhar, L.M., Swanson, N.G., Hurrell, J.J., Ruder, A. and Deddens, J. (1998) Stressors and adverse outcomes for female construction workers. *Journal of Occupational Health Psychology*, 3(1), 19–32.

Goldenhar, L.M., Williams, L.J. and Swanson, N.G. (2003) Modelling relationships between job stressors and injury and near-miss outcomes for construction labourers. *Work and Stress*, 17(3), 218–240.

Gunning, J.G. and Keaveney, M. (1998) A transverse examination of occupational stress among a cross-disciplinary population of Irish construction professionals. In W. Hughes, (ed.), *14th Annual ARCOM Conference*, 9–11 September, University of Reading. Association of Researchers in Construction Management, 1, 98–106.

Haagsma, J.A., Polinder, S., Toet, H., Panneman, M., Havelaar, A.H., Bonsel, G.J. and van Beeck, E.F. (2011) Beyond the neglect of psychological consequences: Post-traumatic stress disorder increase the non-fatal burden of injury by more than 50%. *Injury Prevention*, 17(1), 21–26.

Hans, P.S. (1984) *Construction Management and P.W.D Accounts*, 6th edn. Ludhiana, India: Hans Publications.

Haynes, N.S. and Love, P.E.D. (2004) Psychological adjustment and coping among construction project managers. *Construction Management and Economics*, 22(2), 129–140.

Health and Safety Executive (2007) *An Analysis of the Prevalence and Distribution of Stress in the Construction Industry*. London: Crown (RR518).

Hewage, K.N., Gannoruwa, A. and Ruwanpura, J.Y. (2011) Current status of factors leading to team performance of on-site construction professional in Alberta building construction projects. *Canadian Journal of Civil Engineering*, 38(6), 679–689.

Hibbert L. (2003) Feeling the pressure. *Professional Engineering*, 16(1), 25–26.

Ibem, E.O., Anosike, M.N., Azuh, D.E. and Mosaku, T.O. (2011) Work stress among professionals in the building construction industry in Nigeria. *Australasian Journal of Construction Economics and Building*, 11(3), 46–57.

International Labour Organization (ILO) (1992) Preventing stress at work. In V. Di Martino (ed.), *Conditions of Work Digest*. Geneva: ILO.

Jackson, B.J. (2004) *Construction Management Jumpstart*. Alameda, CA: Sybex.

Janssen, P.P.M., Bakker, A.B. and de Jong, A. (2001) A test and refinement of the demand–control–support model in the construction industry. *International Journal of Stress Management*, 8(4), 315–332.

Kabat-Zinn, J., Massion, A.O., Kristeller, J., Peterson, L.G., Fletcher, K.E., Pbert, L., Lenderking, W.R. and Santorelli, S.F. (1992) Effectiveness of a mediation-based stress reduction program in the treatment of anxiety disorders. *American Journal of Psychiatry*, 149(7), 936–943.

Kennedy, P., Morrison, A. and Milne, D.O. (1997) Resolution of disputes arising from set-off clauses between main contractors and subcontractors. *Construction Management and Economics*, 15(6), 527–537.

Kirmani, S. and Baum, W. (1992) *The Consulting Professions in Developing Countries: A Strategy for Development*. Washington, DC: The World Bank.

Langan-Fox, J. and Cooper, C.L. (2011) *Handbook of Stress in the Occupations*. Oxford: Edward Elgar.

Lazarus, R.S. and Folkman, S. (1984) *Stress, Appraisal and Coping*. New York: Springer.

Leung, M.Y., Wong, M.K. and Oloke, D. (2003) Coping behaviour of construction estimators in stress management. In *Proceedings of 2003 Conference of the Association of Research in Construction Management*, 3–5 September. UK: ARCOM, 1, 271–277.

Leung, M.Y., Ng, S.T., Skitmore, M. and Cheung, S.O. (2005a) Critical stressors influencing construction estimators in Hong Kong. *Construction Management and Economics*, 23(1), 33–43.

Leung, M.Y., Olomolaiye, P., Chong, A. and Lam, C.C.Y. (2005b) Impacts of stress on estimation performance in Hong Kong. *Construction Management and Economics*, 23(9), 891–903.

Leung, M.Y., Liu, A.M.M. and Wong, M.M.K. (2006) Impact of stress-coping behaviour on estimation performance. *Construction Management and Economics*, 24(1), 55–67.

Leung, M.Y., Sham, J. and Chan, Y.S. (2007) Adjusting stressors-job demand stress in preventing rustout/burnout in estimators. *Surveying and Built Environment*, 18(1), 17–26.

Leung, M.Y., Skitmore, M. and Chan, Y.S. (2007) Subjective and objective stress in construction cost estimation. *Construction Management and Economics*, 25(10), 1063–1075.

Leung, M.Y., Chan, Y.S., Chong, A. and Sham, J.F.C. (2008) Developing structural integrated stressor-stress models for clients' and contractors' cost engineers. *Journal of Construction Engineering and Management*, 134(8), 635–643.

Leung, M.Y., Chan, Y.S. and Olomolaiye, P. (2008) Impact of stress on the performance of construction project managers. *Journal of Construction Engineering and Management*, 134(8), 644–665.

Leung, M.Y., Zhang, H. and Skitmore, M. (2008) Effects of organisational supports on the stress of construction estimation participants. *Journal of Construction Engineering and Management*, 134(2), 84–93.

Leung, M.Y., Chan, Y.S. and Yu, J.Y. (2009) Integrated model for the stressors and stresses of construction project managers in Hong Kong. *Journal of Construction and Engineering Management*, 135(2), 126–134.

Leung, M.Y., Chan, Y.S. and Yuen, K.W. (2010) Impacts of stressors and stress on the injury incidents of construction workers in Hong Kong. *Journal of Construction Engineering and Management*, 136 (10), 1093–1103.

Leung, M.Y., Chan, Y.S. and Chong, A.M.L. (2010) Chinese values and stressors of construction professionals in Hong Kong. *Journal of Construction Engineering and Management*, 136(12), 1289–1298.

Leung, M.Y., Chan, Y.S.I. and Dongyu, C. (2011) Structural linear relationships between job stress, burnout, physiological stress and performance of construction project managers. *Engineering, Construction and Architectural Management*, 18(3), 312–328.

Leung, M.Y. and Chen, D. (2011) Exploring the influence of commitment on stress for cost estimators in Hong Kong. *Procedia Engineering*, 14, 1953–1958.

Leung, M.Y. and Chan, I.Y.S. (2012) Exploring stressors of Hong Kong expatriate construction professionals in Mainland China: Focus group study. *Journal of Construction Engineering and Management*, 138(1), 77–88.

Leung, M.Y., Chan, I.Y.S. and Yu, J. (2012) Preventing construction worker injury incidents through the management of personal stress and organisational stressors. *Accident Analysis and Prevention*, 48, 156–166.

Lingard, H. (2003) The impact of individual and job characteristics on 'burnout' among civil engineers in Australia and the implications for employee turnover. *Construction Management and Economics*, 21(1), 69–80.

Lingard, H. (2004) Work and family sources of burnout in the Australian engineering profession: A comparison of respondents in dual and single earner couples, parents and non-parents. *Journal of Construction Engineering and Management*, 130(2), 290–298.

Lingard, H. and Francis, V. (2004) The work-life experiences of office and site-based employees in the Australia construction industry. *Construction Management and Economics*, 22(9), 991–1002.

Lingard, H. and Francis, V. (2005) Does work–family conflict mediate the relationship between job schedule demands and burnout in male construction professionals and managers? *Construction Management and Economics*, 23 (7), 733–745.

Lingard, H. and Francis, V. (2006) Does a supportive work environment moderate the relationship between work-family conflict and burnout among construction professionals? *Construction Management and Economics*, 24(2), 185–196.

Lingard, H., Francis, V. and Turner, M. (2012) Work time demands, work time control and supervisor support in the Australian construction industry: An analysis of work–family interaction. *Engineering, Construction and Architectural Management*, 19(6), 647–665.

Loosemore, M. and Waters, T. (2004) Gender differences in occupational stress among professionals in the construction industry. *Journal of Management in Engineering*, 20(3), 126–132.

Love, P. and Edwards, D.J. (2005) Taking the pulse of UK construction project managers' health: Influence of job demands, job control and social support on psychological wellbeing. *Engineering, Construction and Architectural Management*, 12(1), 88–101.

Love, P.E.D., Edwards, D.J. and Irani, Z. (2010) Work stress, support and mental health in construction. *Journal of Construction Engineering and Management*, 136(6), 650–658.

Lupton, S. (2001) *Architect's Handbook of Practice Management*. London: RIBA Publishing.

Manik, J.A., Greenhouse, S. and Yardley, J. (2013) *Western Firms Feel Pressure as Toll Rises in Bangladesh*. Retrieved from http://www.nytimes.com/2013/04/26/world/asia/bangladeshi-collapse-kills-many-garment-workers.html?pagewanted=all&_r=1&

Martin, J. (2004) Databases for elemental cost planning. *AACE International Transcations*, 02.1–02.5.

Meliá, J. and Becerril, M. (2007) Psychosocial sources of stress and burnout in the construction sector: A structural equation model. *Psicothema*, 19(4), 679–686.

Mitropoulos, P. and Memarian, B. (2012) Team processes and safety workers: Cognitive, affective and behavioural processes of construction crews. *Journal of Construction Engineering and Management*, 138 (10), 1181–1191.

Moorhead, G. and Griffin, R.W. (1995) *Organisational Behavior: Managing People and Organisations*, 4th edn. Boston: Houghton Mifflin.

Morton, R. (2002) *Construction UK: Introduction to the Industry*. Oxford: Blackwell.

Mostert, K., Peeters, M. and Rost, I. (2011) Work–home interference and the relationship with job characteristics and well-being: A South African study among employees in the construction industry. *Stress and Health*, 27(3), e238–e251.

Mow, A.T. (2006) Dispute resolution principles for architects. *Construction Specifier*, 59(7), 26–28.

Muscroft, S. (2005) *Plumbing*, 2nd edn. Oxford: Elsevier.

Ng, S.T. and Chow, L.K. (2004) Framework for evaluating the performance of engineering consultants. *Journal of Professional Issues in Engineering Education and Practice*, 130(4), 280–288.

Ng, S.T., Skitmore, R.M. and Leung, T.K.C. (2005) Manageability of stress among construction project participants. *Engineering, Construction and Architectural Management*, 12(3), 264–282.

Nitithamyong, P. and Tan, Z. (2007) Determinants for effective performance of external project management consultants in Malaysia. *Engineering, Construction and Architectural Management*, 14(5), 463–478.

NYSDOT (2003) *Construction Consultant Manual MURK Part 1D*. New York: New York State Department of Transportation.

Ogunlana, S. (1999) *Profitable Partnering in Construction Procurement*. London: E. & F.N. Spon.

Plunkett, J.W. (2007) *Plunkett's Real Estate Construction Industry Almanac*. Plunkett Research.

PMI (2000) *A Guide to the Project Management Body of Knowledge*. Pennsylvania: Project Management Institute.

Poon, S.W., Rowlinson, S.M., Koh, T. and Deng, Y. (2013) Job burnout and safety performance in the Hong Kong construction industry. *International Journal of Construction Management*, 13(1), 69–78.

Psilander, K. (2007) Why are small developers more efficient than large developers? *Journal of Real Estate Portfolio Management*, 13(3), 257–267.

Quatman, G.W. (2000) *Design-Build for the Design Professional*. New York: Aspen.

Oyedele, L.O. and Tham, K.W. (2007) Clients' assessment of architects' performance in building delivery process: Evidence from Nigeria. *Building and Environment*, 42(5), 2090–2099.

RICS (1976) *An introduction to Cost Planning*. London: RICS.

RICS (2011a) *Defining Completion of Construction Works – RICS Guidance Note*. Retrieved from http://www.joinricsineurope.eu/uploads/files/Part7RICSDefining completiononconstructionworksBlackBook_1.pdf

RICS (2011) *Real Estate and Construction Professionals in India by 2020*. London: RICS.

Rowley, C. and Jackson, K. (2011) *Human Resource Management: The Key Concepts*. London: Routledge.

Sang, K.J.C., Dainty, A.R.J. and Ison, S.G. (2007) Gender: A risk factor for occupational stress in the architectural profession? *Construction Management and Economics*, 25, 1305–1317.

Schwartz, M. (1993) *Basic Engineering for Builders*. New York: Craftsman.

Sears, S.K., Sears, G.A. and Clough, R.H. (2008) *Construction Project Management: A Practical Guide to Field Construction Management*. Chichester: Wiley.

Shoesmith, D.R. (1996) A study of the management and procurement of building services work. *Construction Management and Economics*, 14, 93–101.

Siu, O.L., Philips, D.R. and Leung, T.W. (2004) Safety climate and safety performance among construction workers in Hong Kong: The role of psychological strains as mediators. *Accident Analysis and Prevention*, 36(3), 359–366.

Smith, J. and Jaggar, D. (2007) *Building Cost Planning for the Design Team*, 2nd edn. Burlington: Elsevier.

Sommerville, J. and Langford, V. (1994) Multivariate influences on the people side of projects: Stress and conflict. *International Journal of Project Management*, 12(4), 234–243.

Stein, S.G.M. and Hiss, R. (2003) Here comes the judge – Duties and responsibilities of design professionals when deciding disputes. *Journal of Professional Issues in Engineering Education and Practice*, 129(3), 177–183.

Sutherland, V.J. and Davidson, M.J. (1989) Stress among construction site managers: A preliminary study. *Stress Medicine*, 5, 221–235.

Sutherland, V. and Davidson, M.J. (1993) Using a stress audit: The construction site manager experience in the UK. *Work Stress*, 7, 236–286.

Tabassi, A.A. and Bakar, A.H.A. (2009) Training, motivation and performance: The case of human resource management in construction projects in Mashhad, Iran. *International Journal of Project Management*, 27, 471–480.

Telgen, J. and Schotanus, F. (2010) Effects of full transparency in supplier selection on subjectivity and bid quality. *The 4th International Public Procurement Conference*. August 26–28, Seoul.

Tordoir, P.P. (1995) *The Professional Knowledge Economy: The Management and Integration Services in Business Organisations*. Dordrecht: Kluwer.

Turner, J.R. (1999) *The Handbook of Project-based Management*, 3rd edn. New York: McGraw-Hill.

Vocational Training Council and Building and Civil Engineering Traing Board (2011) *2011 Manpower Survey Report-Building and Civil Engineering Indsutry*. Hong Kong.

Wallace, S. (2007) *The ePMbook*. Retrieved at http://www.epmbook.com/, viewed on 08 September, 2013.

Wallis, J.J. and Oaates, W.E. (1998) *Decentralization in the Public Sector: An Empirical Study of State and Local Government*. Retrieved from http://www.nber.org/chapters/c7882.pdf

Waters, E.A. (2004) Assessing consultant performance: Could do better? *Clinician in Management*, 12, 83–87.

Weatherley, E. and Irit, P. (1996) Coping with stress: Public welfare supervisors doing their best. *Human Relations*, 49(2), 157–170.

Wells, J. (1986) *The Construction Industry in Developing Countries: Alternative Strategies for Development*. Beckenham: Croom Helm.Winch, G.M. (2001) Governing the project process: A conceptual framework. *Construction Management and Economics*, 19, 799–808.

Winch, G.M. (2010) *Managing Construction Projects*, 2nd edn. Oxford: Wiley-Blackwell.

Yerkes, R.M. and Dodson, J.D. (1908) The relation of strength of stimulus to rapidity of habit formation. *Journal of Comparative Neurology and Psychology*, 18, 459–482.

Yip, B., Rowlinson, S. and Siu, O.L. (2008) Coping strategies as moderators in the relationship between role overload and burnout. *Construction Management and Economics*, 26(8), 869–880.

Yip, B. and Rowlinson, S. (2009a) Job redesign as n intervention strategy of burnout: Organisational perspective. *Journal of Construction Engineering and Management*, 135(8), 737–745.

Yip, B. and Rowlinson, S. (2009b) Job burnout among construction engineers working within consulting and contracting organisations. *Journal of Management in Engineering*, 25(3), 122–130.

Zhou, Q., Fang, D. and Wang, X. (2008) A method to identify strategies for the improvement of human safety behavior by considering safety climate and personal experience. *Safety Science*, 46(10), 1406–1419.

2

Theories of Stress

Before looking into the various theories of stress, it is worthwhile to review how the concept of stress itself has evolved. The idea was initially proposed by a group of psychobiologists in the 1920s, while the term 'stress' was first used in the early twentieth century to denote a general anxiety rooted in the cares of life (Cannon 1914). The concept has been applied and investigated in different contexts by psychobiologists (focusing on the minds and behaviours of individuals: Langeland and Olff 2008), sociologists (focusing on groups: Pearlin 1989) and psychiatrists (focusing on diagnosis, treatment and prevention: Puca et al. 1999). Alongside this strand of research, various theories have been developed to define various stress-related variables, specify domains, establish relationships and predict specific aspects (Hunt 1991; Reynolds 1971; Wacker 1998). This chapter discusses the history of these stress theories as promulgated by various schools of thought and gives an overview of the various arousal, appraisal and regulatory models of stress.

2.1   The History of Stress Theory

Although the term stress was first used in the 1920s, it was not generally adopted in the popular literature until the early 1950s when Hans Selye embarked on his research (Selye 1955). At that time, stress was regarded as a general concept describing the difficulties and/or demands facing an individual arising from general life circumstances, leading to psychological and/or biological disease. In fact, the term stress originated from the Romantic critique of modernity in the nineteenth century, which proposed that stress (or strain) was a cause of mental disease (Rosenberg 1962). Due to these

Stress Management in the Construction Industry, First Edition.
Mei-yung Leung, Isabelle Yee Shan Chan and Cary L. Cooper.
© 2015 John Wiley & Sons, Ltd. Published 2015 by John Wiley & Sons, Ltd.

aetiological theories that considered stress to be the cause of illness and the trend towards 'moral therapy' to alleviate stress, the first mental hospitals were built as part of the American Revolution during the Jacksonian era (Abbott 1990). Among the various moral causes of mental ill-health, such as work overload and marital problems, the pressure of an urban, industrial and commercial civilisation that was considered to be unnatural to the human organism was regarded as the main element (Grob 1973). In the late nineteenth century, the causes of mental disease identified in the Jacksonian era were considered to be the antecedents of nervous disease resulting in neurasthenia, which is similar to the syndrome of anxious depression identified in more recent literature. At that time, researchers identified stress as an antecedent of neurasthenia, as it is usually followed by minor heart pain, headaches and other nervous symptoms (Abbott 1990). This was the start of how psychobiologists came to associate stress with the physiological problems suffered by human beings.

2.1.1   Psychobiology

Psychobiology refers to a school of thought that focuses on one's mind and behaviour. Since psychobiologists in the twenty-first century now incorporate both proximate and ultimate approaches to understanding behaviour, the 'biology' in psychobiology should be regarded as forming part of the holistic approaches of ethology, ecology, evolution and comparative psychology, as well as drawing on the latest physiological methods and thought (Dewsbury 1991).

The early psychobiologists identified a connection between emotional stress and physical well-being. Cannon (1927) carried out some early experiments showing that the stimuli associated with emotional arousal led to changes in physiological processes. The first studies on the response of the organism to stressful situations in general and on the psychobiology of stress in particular, are probably those of Cannon and De La Paz (1911). They investigated the involvement of the adrenal medulla and the sympathetic system in emergency situations (Cannon 1927). Cannon noted that the venous blood of cats frightened by barking dogs contained adrenaline, a response of the organism which can be prevented by adrenalectomy or by sectioning the splanchnic nerve innervating the adrenal medulla. This suggested that the adrenal medulla was acting in concert with the sympathetic nervous system such that both systems were activated during stress (Puglisi-Allegra 1990). Cannon conducted further work on the fight or flight response theory in the 1930s. This theory is advanced to explain the increased activity of the sympathetic nervous system to produce the necessary alertness and energy required for action (such as increased arousal and vigilance, increase in heart rate and contraction force of the heart, enhanced blood pressure, dilation of the bronchi, redistribution of the blood from the internal organs to the muscles and increased gluconeogenesis) (Vingerhoets and Perski 2000).

Later on, Hans Selye developed the notion of the general adaptation syndrome (Selye 1976c). This nonspecific response of the body to any demand has three characteristics, including adrenocortical enlargement, thymicolymphatic involution and intestinal ulcers. The syndrome is characterised as having three phases: alarm reaction, resistance, and exhaustion. In the final phase, there is a total breakdown of the organism and a complete loss of resistance gives way to the development of what are called 'diseases of adaptation'. In fact, studies of hormonal reactions to stress stimulation often give rise to paradoxical findings. An increased release of a particular hormone may be observed in some situations, whereas in others it might be suppressed. Theories developed by psychobiologists tend to propose that it is the individual's interpretation or appraisal of the stressful situation, rather than its objective characteristics, that determines the nature of the endocrine and cardiovascular responses (Gaab et al. 2005). When facing similar situations, different individuals may experience eustress or distress due to their different personal characteristics and experiences (Selye 1978; please refer to Section 2.2.2 for details of eustress and distress). In addition, the biological status of the individual (such as phase of the menstrual cycle, pregnancy, smoking status and physical fitness), genetic factors, age and social support are also relevant as codeterminants of the biological stress response (Vingerhoets and Perski 2000). This sheds light on the subsequent development of the person–environment (P–E) fit model (Sherry 1991; please refer to Section 2.1.2 for more details of this theory).

2.1.2   Sociology

Based on the knowledge developed by the psychobiologists, sociologists then began to research stress (Weber 2010). Unlike psychobiology, which focuses on the minds and behaviours of individuals, sociology emphasises the development, structure and functioning of human society (Barbalet 1998). The first sociologist to conduct stress-related research was Lindemann (1944), who investigated the individual experiences of bereavement of the surviving relatives of those who died in the Melody Lounge Cocoanut Grove fire in Boston. He observed people who had positive outcomes in a grieving process. Examples of pathognomonic signs included somatic distress, preoccupation with the image of the deceased, guilt, hostile reactions, loss of patterns of conduct and the appearance of traits of the deceased. Lindemann's work acted as a platform for the development of the discipline of crisis management – a term now used interchangeably with stress management.

In addition to individual grief reactions, many stressful experiences are related to the social structure or environment of an individual, such as social and economic class, race, gender and age, which determine the different levels of power and resources available to an individual. Being at low levels of these social structures may in itself be a stressor. In addition, such roles and statuses often persist over time, with long-term influences on an individual. The results of social studies conducted in the 1950s and 1960s generally indicate significant relationships between the social environment and individual well-being (Pearlin 1989).

Following these studies, French and colleagues, as social studies researchers, started to develop the P–E fit theory (French and Kahn 1962; French, Rodgers and Cobb 1974). According to this theory, stress is the result of a misfit between the person and his/her environment. There are three basic distinctions which are core to the theory: (i) the person versus the environment; (ii) the objective versus the subjective representations of each of these; and (iii) between the two types of fit, namely that between environmental demands and individual abilities and between the needs of the person and the environmental supplies that are suitable to meet those needs. Based on this theory, stress arises when there is a misfit between what is being offered in the environment and the needs of the person and/or if someone lacks the ability to meet the demands made of him/her in order to receive supplies. In other words, a misfit between the abilities and demands made of an individual does not in itself cause stress; this only occurs when there are excessive demands which must be met in order for the individual to receive supplies. The P–E fit theory has been further developed and refined by many different researchers (see e.g. Caplan 1983, 1987; Edwards, 1996; Edwards and Cooper 1990, van Harrison, 1978, 1985). It will be discussed further in Section 3.1.1.

2.1.3   Psychiatry

After the sociologists had explored individual reactions to stressful social environments, psychiatrists started researching stress from different perspectives, including not only how individuals react, but also the sources of stress, the assessment of stress and stressors and so on. Psychiatrists tend to regard stress as an illness and to focus on its causes, diagnosis, treatment and prevention (Tennant 1994). Tyhurst (1951, 1957a, 1957b) first embarked on the investigation and development of a model of individual reactions to disaster. It has three stages: impact, recoil, and the posttraumatic period. There are different degrees or levels of stress, duration and psychological and social status across these three stages. When the initial disaster occurs, the individual suffers from the maximum and direct effect of it. The period of recoil denotes the period when the initial stress is no longer present and the individual has begun to cope. Although the coping behaviours adopted may be negative (such as avoidance or escapism) and fail to eliminate the stress altogether, an individual may still subjectively feel that its impact has been alleviated. Lastly, the stress of the posttraumatic period is more closely related to social factors. After an individual has come to fully understand the impact of the disaster, such as the loss of home or of family members, he/she faces a different life in a changed environment.

Caplan (1964, 1974) subsequently developed the stage theory of crisis development in order to investigate how some individuals avoid mental illness. This three-stage model was developed to prevent the need for psychiatric intervention and enhance mental health and includes primary (avoiding mental health incidents; that is, stressors), secondary (identification and treatment of mental health problems at an early stage), and tertiary (moderating the outcomes of mental health problems) forms of prevention. Caplan's model has been widely adopted by researchers in diverse fields such as workplace stress in an organisational setting (see e.g. Quick et al. 1997), behavioural and emotional disorders of children in a school setting (Offord 1987) and public health (Bowen and Muhajarine 2010).

In addition to the process and three-stage prevention models, Holmes and Rahe (1967), building on the work of Tyhurst, developed the Social Readjustment Rating Scale (SRRS) for the assessment of stress and stress management. The scale has been widely adopted, revised and updated by researchers such as Hobson et al. (1998) and De Coteau, Hope and Anderson (2003). Moreover, another stress scale developed by Holmes and Rahe, comprising a list of 43 stressful life events which contribute to illness and require a change in the normal life pattern of individuals, is now a cornerstone of the assessment of vulnerability to stress (see e.g. Lewis et al. 2003).

2.2   Arousal Theories

Research into the early stages of stress tends to focus on arousal, or any physical and psychological activity which affects the performance of an individual. Arousal level is a hypothetical construct giving a general indication of the stimulation level of an individual, such as heart rate and blood pressure. In general, arousal theories emphasise stressors (stimulus), stress reactions (such as physical and psychological signs) and the relationship between them.

2.2.1   Fight or Flight – Cannon

The fight or flight response theory was first developed by Cannon (1927) who conducted empirical studies on emotional glycosuria based on McDougall's (1908) work on the emotions and reactions of the individual. It focuses on the reactions of individuals to different emotions. For instance, anger is associated with fighting, while fear is associated with flight (Cannon 1914).

In the past, when our ancestors were living in the wild and hunting for food, the fight or flight response system was essential to survival. When danger arises, hormones are released from the brain in order to support other parts of the body to deal with the circumstances, such as fighting the animal or fleeing to escape from it. After a certain period of time, these physical adjustments can revert to normal when the individual is no longer threatened by the stressor. However, modern humans no longer live in the wild and the stressors facing them are usually chronic in nature, such as work overload due to task complexity and/or delay in information processing. The fight or flight response theory has been used as a metaphor for human behavioural responses to stress and emotion and whether one fights or flees in response to sympathetic arousal is considered to depend on the nature of the stressor. Although stressors can be objective, perceptions of individuals can be highly subjective. If an individual considers a stressor to be of limited influence and perceives that he/she has a chance at overcoming it, then the likely outcome is a fight (Dewe and Cooper 2012). In contrast, if an individual perceives a stressor as frightening, flight is a more likely choice (Taylor et al. 2000).

In the 'black box' between individual emotions and reactions, Cannon (1927) identifies various biological adjustments, including rapid respiration, increase in blood sugar, secretion of adrenalin and altered circulation through pain and emotional excitement, which support the body to either struggle or flee. In other words, the fight or flight response system, through various physiological adjustments, serves to support an individual's actions and/or behaviours in coping with stress, such as escapism (see Chapter 6 for more information about coping behaviours).

2.2.2   Eustress versus Distress – Selye

Physiological responses to stressors are general in character, since the set of possibilities is similar across different stressors and contexts. The term stress was first developed to describe a negative physiological state caused by adverse situations (Selye 1964). However, this approach ignores an important aspect of stress, namely the *positive* physiological response of an individual in pleasant circumstances. Based on a series of studies on the physiological response to stressors, Selye (1978) defines stress as 'the non-specific response of the body to any demand, whether it is caused by, or results in, pleasant or unpleasant conditions' (p. 13). He goes on to develop the concept of eustress versus distress. Eustress (from the Greek *eu* meaning good, as in euphonia or euphoria) and distress (from the Latin *dis* meaning bad, as in dissonance or disagreement) are different and distinct from each other. Distress occurs when the demands placed on an individual exceed his/her ability to cope. In other circumstances, distress may also occur when such demands are far *lower* than his/her ability, because of the need of an individual for accomplishment. Following this logic, eustress can be considered as a moderate (optimal) level of stress (refer to Section 5.5.2.2 for more information about optimisation of stress) which is a good fit with external demands.

However, this is only one aspect of the concept of eustress versus distress. Although demands are objective events or situations, different people will perceive them subjectively as pleasant or unpleasant, due to their different abilities and life experiences. Selye (1987) therefore proposes that the distressing or eustressing nature of a potential stressor is determined by the individual's interpretation of it and his/her chosen response. Hence, although previous researchers interpret eustress differently (e.g. Harris (1970) refers to it as pleasure, while Edwards and Cooper (1988) see it as a discrepancy between perceptions and desires), all agree that eustress depends on the perception of the individual (Le Fevre, Matheny and Kolt 2003).

The eustress versus distress system was a breakthrough in the development of the concept of stress in the academic and, more importantly, the medical fields. Instead of recommending drugs to eliminate stress in all circumstances, a careful evaluation should be conducted to ensure medication is used only in cases of distress, not eustress (Selye 1976a). A case of eustress, such as a work environment which is challenging but within the capacity of an individual professional to handle, generates feelings of satisfaction when work tasks are accomplished and can be considered important to the development of that individual. Studies indicate eustress and distress can result in different adrenal corticosteroids being generated, with eustress having a positive impact on an individual by means such as enhancing the immune system, while distress may impair it (Lazarus 1993).

2.2.3   The Yerkes–Dodson Law

The Yerkes–Dodson law was developed based on an empirical study of the relationship between the strength of a stimulus and the rate of learning conducted by Yerkes and Dodson (1908). Mice were the subjects of the experiment. They were presented with two boxes, a white one and a black one and required to choose the former regardless of their relative positions. If the mice attempted to enter the black box, they received a disagreeable electric shock. The results show that the relationship between the strength of a stimulus and the rate of learning, or habit formation, depends on the difficulty of the habit to be learned (in this experiment, the condition of visual discrimination). For learning with a low level of difficulty (e.g. when the boxes can be easily distinguished because one is much brighter than the other), the speed of learning increases with the increase in the strength of the electrical stimulus. For more difficult learning (such as the situation where the boxes differ only slightly in brightness, making it harder for the mice to learn how to discriminate between them), the speed of learning initially increases with the increase of the strength of the stimulus. However, beyond a certain level of intensity of stimulation, which is quickly reached, the rate of learning starts to decrease. In this case, both weak and strong stimuli result in a low rate of learning. There is a threshold point for the stimulus which results in the highest rate. Then, for even more difficult learning (i.e. the most difficult condition in this experiment), similar results are obtained to the moderately difficult case even though the threshold point is lower. In other words, more difficult learning may be readily acquired only under relatively weak stimulation. These results help to explain a number of studies which have investigated the inverted U-shaped relationship between stress and performance in general and for construction personnel in particular (refer to Section 5.2.2.2 for the optimisation or threshold point of stress).

Based on this work, the Yerkes–Dodson law has been further adapted to investigate or explain the influence of punishment, reward, motivation, drive, arousal, anxiety, performance and so on. In this body of work, visual discrimination is also commonly referred to as the difficulty, complexity or novelty of various tasks (Teigen 1994). Researchers have adopted the law in stress studies and demonstrated empirically that the threshold level of the stress–performance relationship is not fixed. Optimum levels vary across individuals and tasks, while different tasks may require different levels of

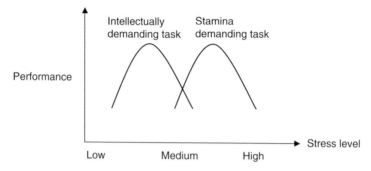

Figure 2.1   Relationship between performance and stress for different tasks.

arousal (Meglino 1977; Selye 1976b). For example, difficult or intellectually demanding tasks may require a lower level of arousal to facilitate concentration for optimal performance, whereas tasks demanding more stamina or persistence may be performed better at higher levels of arousal (to increase motivation). Therefore, a more precise model emerges (see Figure 2.1). Some researchers predict the relationship between stress and performance to be monotonically positive for simple and well-learned tasks, with the inverted U-shaped relationship only predicted for complex tasks involving many cues. It has been argued that given a change in direction after the optimum, high stress causes cognitive narrowing and rigidity of behaviour. This seems particularly applicable to complex tasks where novel responses, problem solving, or attention to many task elements will be necessary.

2.3   Appraisal and Regulatory Theories

The development of the stress management research has seen the focus expanded to cover, explicitly or implicitly, a cognitive appraisal (evaluation of the environment) and self-regulation (selection of coping strategies) mechanism within the process. Instead of investigating how an individual is influenced by stressors, or the relationships between stressors and stress, these later theories focus on how the individual appraises various situations and how this influences his/her selection of appropriate responses (i.e., coping).

2.3.1   Transactional Stress Model – Lazarus and Folkman

The transactional stress model developed by Lazarus and Folkman (1984) aims to evaluate the processes of coping with stressors. It proposes that stress and its appraisal are related to neither the person nor the environment alone. For example, two construction workers operating at the same level of height (i.e., in the same environment) may not necessarily experience the same degree of stress in response. Individuals appraise the same environment differently, due to their different experiences and psychological characteristics. A worker who suffers from acrophobia or who has previously witnessed a

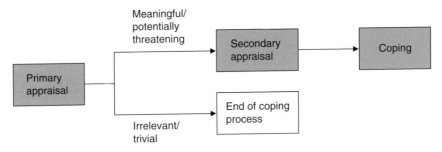

Figure 2.2   Three stages of the coping process.

death from falling could become tremendously stressed about working at height compared to one who has not had these experiences. However, the same worker could be considerably less worried about a risky situation at ground level. In other words, appraisal is a transactional variable which depends on the interaction between the environment and the person.

To take a step further, neither the environment nor the person are static. They are constantly changing variables, as are their interactions. As the environment changes, so too does the perception of the person. Continuing the previous example, the worker who is afraid of working at heights may feel less stressed about this after the façade of the building has been completed, since he/she will then be operating within the internal building structure which may no longer feel like working at height. The shift of emotions reflects changes in the meaning of the P–E interaction as it changes over time. Hence, stress is a product of the interplay between the environment and the person at any given moment. By appraising different situations (potential stressors), an individual can actively select and shape the environment through appraisal and selectivity. There are three stages of appraisal in the coping process (Lazarus and Folkman 1984): (i) primary appraisal; (ii) secondary appraisal; and (iii) coping (see Figure 2.2).

2.3.1.1 *Primary appraisal*

In this stage, individuals decide whether they are potentially threatened or placed in jeopardy by a situation. In other words, they appraise whether or not it is worth worrying about. If the situation is judged to be irrelevant or trivial, the coping process ends here. In contrast, if the circumstance is considered meaningful and potentially threatening, it continues to the next phase (Lazarus and Folkman 1984).

2.3.1.2 *Secondary appraisal*

The individual then assesses his/her resources for dealing with the stressors: 'This assessment is influenced by previous experiences in similar situations, generalised beliefs about the self and the environment and the availability of personal and environmental resources' (Holroyd and Lazarus 1982). The assessment of how much control one has over a situation is also important at this stage. The lesser the perceived control, the more threatening the situation and the greater probability of mental and physical distress as a result (Schafer 2000).

2.3.1.3 *Coping*

In this stage, individuals take whatever actions they regard as appropriate. These might include action of some kind, or a cognitive adjustment. However, the effectiveness of coping response of an individual is another matter (Frese 1986; Krohne 1986; Laux 1986; Schafer 2000). The details of coping will be discussed in Chapter 6.

Based on the above model, Lazarus and Folkman (1984) define psychological stress as the outcome of an individual's appraisal of his/her environment as exceeding his/her ability to cope with it.

2.3.2   Model of Controlled Processing and Compensatory Control – Hockey

Behaviour is a goal-directed activity controlled by self-regulatory processes which involve effortful analyses of performance and cost. Hockey (1997) distinguishes effort as controlled processing (such as mental resources) from effort as compensatory control (such as arousal level) and proposes a cognitive–energy framework to show that maintaining stable performance under stressful conditions is an active process involving effort and resources. This requires controlled processing such as working memory capacity and the management of cognitive resources via mental effort. In order to control the effectiveness of various behaviours, an individual manages his/her efforts in relation to different environments and energy levels. Hockey introduces the notion of 'performance protection', in which an individual will seek to maintain high-priority task goals under acceptable limits by either spending extra cost (i.e., committing extra resources) or accepting a reduction in performance to avoid such expenditure. In most situations, an increased commitment to task goals results in a decrease in the relevance of other personal or biological goals, such as leisure, rest or health.

Hockey (1997) proposes that an individual supervises and manages his/her mental resources and effort based on two feedback regulation mechanisms by comparing his/her current activity and goal-based performance level. There are two possible feedback loops. The first focuses on escalating the degree of effort and/or resources to maintain goal performance. To achieve this for simple tasks (such as domestic activities), an individual tends to exert control at a lower level that demands little effort and resource allocation. For more complicated tasks, he/she employs a supervisory controller at a higher level, with resources being allocated to compensate for the larger discrepancies between the current level of activity and goal-based performance. The second process emphasises altering the goals of tasks and/ or performance so as to maintain low levels of effort, or to reduce effort to prevent prolonged exposure to stress. This is regarded as a passive form of coping.

Although the controlled processing and compensatory control mechanisms provide a conceptualisation of the association between environment, effort and response, it is difficult to test the theory empirically since the concept of effort and resources (such as mental resources or energy) cannot easily be measured or quantified (Parasuraman and Rizzo 2008).

2.3.3   Stress-Adaptation Model – Hancock and Warm

Hancock and Warm (1989) take an input–adaptation–output approach to developing the stress-adaptation model. They assume firstly that individuals invest effort in adapting to stressful environments for the sake of behavioural and physiological stability and secondly that under different levels of stress, adaptability or capability (attentional resources) are drained. They describe environmental stress (equivalent to the concept of the stressor used throughout this book) as input, coping as nomothetic adaptation and psychological and/or physical responses (equivalent to our conceptualisation of stress) as ideographic outputs. The model theorises changes in behavioural adaptation as a function of stress level, in which the relationship between the two is characterised by an inverted U-shaped curve. Based on this model, there are three modes of operation, representing three different relationships among stress, adaptability and response. The concepts of hypostress (understimulation, where the sensory system experiences inadequate levels of stimulation) and hyperstress (overstimulation, or excessive stimulation of the sensory system) are introduced to indicate the boundaries which cause dynamic instability in extreme cases. In the continuum between hypo- and hyperstress, the comfort zone reflects the highest level of adaptability. In other words, this level can be achieved by a minor (neutral) level of stress. However, an increase in the intensity of stress (either hypo- or hyperstress) decreases adaptability. It is worth noting that the stress-adaptation model addresses both physiological (such as homeostatic regulation) and psychological (such as mental resources allocation) adaptations. The comfort zone of the latter is usually smaller than the former, such that physiological adaptation usually follows the depletion of psychological resources.

The model has been further extended to focus on the association between two kinds of task stress, namely information rate (i.e., the temporal flow of information in the environment) and information structure (i.e., the meaning of information as perceived by individuals) on behavioural and physiological adaptability. To acknowledge the interaction between stressors in the real-life environment, a three-dimensional model has been developed in which the stress axis is divided into two dimensions and the relationships of each of these with adaptation are also conceptualised. Since individuals interpret information in various ways, stereotypical behaviours should not be observed at mild levels of stress, where there are many options available for achieving the goals. Common behaviours across individuals are expected under extreme levels of stress, where there are limited solutions (perhaps even only one) and attention narrows (i.e., as stress escalates, a person focuses more closely on the cues perceived as having greatest salience). In order to maximise the capacity for task performance towards a goal, the individual undergoes a process of attention narrowing so as to retain optimum amounts of information by adjusting perceived rate and structure. Similar to the three-dimensional model described above, there is a normative zone reflecting minor levels of stress that are not enough to arouse adaptive actions. In the meantime, there are boundaries for extreme cases, namely hyper- and hypostress. The stress-adaptation model demonstrates the importance of incorporating the interactions between stressors into the stress management process (see Section 4.3.1 for interactions of stressors in the construction context).

2.4   Summary

This chapter has presented a general review of the history of stress theory from the perspective of three disciplines, namely psychobiology, sociology and psychiatry. This was followed by descriptions of various theories, some of which have been developed on a cross-disciplinary basis. The shift from arousal (i.e., fight or flight, eustress versus distress and the Yerkes–Dodson law) to appraisal and regulatory (transactional stress, controlled processing and compensation and stress-adaptation models) theories demonstrates a departure from focusing on stressors (stimulus) and stress reactions (such as physical and psychological signs) in the early days towards an explicit or implicit acknowledgement of appraisal and coping mechanisms in later studies. However, there remain few, if indeed any, comprehensive models or theories covering the entire process of stressors, stress, coping behaviours and performance in general, let alone for the construction industry as a particular setting. This demonstrates the importance of developing a comprehensive stress management model for construction personnel working in the industry.

References

Abbott, A. (1990) Positivism and interpretation in sociology: Lessons for sociologists from the history of stress research. *Sociological Forum*, 5(3), 435–458.

Barbalet, J.M. (1998) *Emotion, Social Theory, and Social Structure: Emotion in Social Life and Social Theory*. Cambridge: Cambridge University Press.

Bowen, A. and Muhajarine, N. (2010) Integrating population health promotion and prevention: A model approach to research and action for vulnerable pregnant women. In T.A. McIntosh, B. Jeffery and N. Muhajarine (eds), *Redistributing Health: New Directions in Population Health Research in Canada*, 133–150. Regina: CPRC Press.

Cannon, W.B. (1914) The interrelations of emotions as suggested by recent physiological researches. *American Journal of Psychology*, 25(2), 256–282.

Cannon, W.B. (1927) Bodily changes in pain, hunger, fear and rage. *Southern Medical Journal*, 22(9), 870.

Cannon, W.B. and De La Paz, D. (1911) The stimulation of adrenal secretion by emotional excitement. *Journal of the American Medical Association*, 1(10), 742.

Caplan, G. (1964) *Principles of Preventive Psychiatry*. New York: Basic Books.

Caplan, G. (1974) *Support Systems and Community Mental Health*. New York: Human Sciences Press.

Caplan, R.D. (1983) Person–environment fit: past, present and future. In C.L. Cooper (ed.), *Stress Research*. New York: Wiley.

Caplan, R.D. (1987) Person–environment fit theory and organizations: Commensurate dimensions, time perspectives and mechanisms. *Journal of Vocational Behaviour*, 31(3), 248–267.

De Coteau, T.J., Hope, D.A. and Anderson, J. (2003) Anxiety, stress and health in Northern Plains Native Americans. *Behaviour Therapy*, 34(3), 365–380.

Dewe, P. and Cooper, C.L. (2012) *Wellbeing and Work: Towards a Balanced Agenda*. Basingstoke: Palgrave Macmillan.

Dewsbury, D.A. (1991) Psychobiology. *American Psychologist*, 46(3), 198–205.

Edwards, J.R. (1996) An examination of competing versions of the person–environment fit approach to stress. *Academy of Management Journal*, 39(2), 292–339.

Edwards, J.R. and Cooper, C.L. (1990) The person–environment fit approach to stress: Recurring problems and some suggested solutions. *Journal of Organizational Behaviour*, 11(4), 293–307.

French, J.R.P. and Kahn, R.L. (1962) A programmatic approach to studying the industrial environment and mental health. *Journal of Social Issues*, 18(3), 1–48.

French, J.R.P, Rodgers, W. and Cobb, S. (1974) Adjustment as person–environment fit. In G.V. Coelho, D.A. Hamburg and J.E. Adams (eds.), *Coping and Adaptation*. New York: Basic Books.

Frese, M. (1986) Coping as a moderator and mediator between stress at work and psychosomatic complaints. In M.H. Appley and R. Trumball (eds), *Dynamic of Stress. Physiological, Psychological, and Social Perspectives*. New York: Plenum.

Gaab, J., Rohleder, N., Nater, U.M. and Ehilert, U. (2005) Psychological determinants of the cortisol stress response: The role of anticipatory cognitive appraisal. *Psychoneuroendocrinology*, 30(6), 599–610.

Grob, G.N. (1973) *Mental Institutions in America: Social Policy to 1875*. New York: Transaction.

Hancock, P.A. and Warm, J. S. (1989) A dynamic model of stress and sustained attention. *Human Factors*, 31(5), 519–537.

Harris, D.V. (1970) On the brink of catastrophe. *Quest*, 13, 33–40.

Holroyd, K.A. and Lazarus, R.S. (1982) Stress, coping and somatic adaptation. In L. Goldberger and S. Breznitz (eds), *Handbook of Stress: Theoretical and Clinical Aspects* (3rd edn). New York: Macmillan.

Hobson, C.J., Kamen, J., Szostek, J., Nethercut, C.M., Tiedmann, J.W. and Wojnarowicz, S. (1998) Stress life events: A revision and update of the social readjustment rating scale. *International Journal of Stress Management*, 5(1), 1–23.

Hockey, G.R.J. (1997) Compensatory control in the regulation of human performance under stress and high workload: A cognitive–energetical framework. *Biological Psychology*, 45(1), 73–93.

Holmes, T.H. and Rahe, R.H. (1967) The social readjustment rating scale. *Journal of Psychosomatic Research*, 11(2), 213–218.

Hunt, S.D. (1991) *Modern Marketing Theory: Critical Issues in the Philosophy of Marketing Science*. Cincinnati, OH: Southwestern Publishing.

Krohne, H. W. (1986) Coping with stress: Dispositions, strategies, and the problem of measurement. In M.H. Appley and R. Trumball (eds), *Dynamics of Stress: Physiological, Psychological, and Social Perspectives*. New York: Plenum Press.

Laux, L. (1986) A self-presentational view of coping with stress. In M.H. Appley and R. Trumball (eds), *Dynamics of Stress: Physiological, Psychological, and Social Perspectives*. New York, Plenum Press.

Langeland, W. and Olff, M. (2008) Psychobiology of posttraumatic stress disorder in pediatric injury patients: A review of literature. *Neuroscience and Biobehavioural Reviews*, 32(1), 161–174.

Lazarus, R. S. (1993) From psychological stress to the emotions: A history of changing outlooks. *Annual Review of Psychology*, 44(1), 1–21.

Lazarus, R.S. and Folkman, S. (1984) *Stress, Appraisal and Coping*. Springer.

Le Fevre, M., Matheny, J. and Kolt, G.S. (2003) Eustress, distress and interpretation in occupational stress. *Journal of Managerial Psychology*, 18(7), 726–744.

Lewis, J.A., Lewis, M.D., Daniels, J.A. and D'Andrea, M.J. (2003) *Community Counseling: Empowerment Strategies for a Diverse Society* (3rd edn). Pacific Grove, CA: Thomson Brooks/Cole.

Lindemann, E. (1944) Symptomatology and management of acute grief. *American Journal of Psychiatry*, 101(2), 141–148.

McDougall, W. (1908) *An Introduction to Social Psychology*. London: Methuen.

Meglino, B.M. (1977) The stress-performance controversy. *MSU Business Topics*, 25, 53–59.

Offord, D.R. (1987) Prevention of behavioural and emotional disorders in children. *Journal of Child Psychology and Psychiatry*, 28(1), 9–19.

Parasuraman, R. and Rizzo, M. (2008) *Neuroergonomics: The Brain at Work*. Oxford: Oxford University Press.

Pearlin, L.I. (1989) The sociological study of stress. *Journal of Health and Social Behaviour*, 30(3), 241–256.

Puca, F., Genco, S., Prudenzano, M.P., Savarese, M., Bussone, G., Amico, D., Cerbo, R., . . ., Marabini, S. (1999) Psychiatric comorbidity and psychosocial stress in patients with tension-type headache from headache centers in Italy. *Cephalalgia*, 19(3), 159–164.

Puglisi-Allegra, S. (1990) Psychobiology of stress. *Behavioural and Social Sciences*, 54.

Quick, J.C., Quick, J.D., Nelson, D.L. and Hurrell, J.J. (1997) *Preventive Stress Management in Organizations*, Washington, DC: American Psychological Association.

Reynolds, P.D. (1971) *A Primer in Theory Construction*. Indiana: Bobbs-Merrill.

Rosenberg, C.E. (1962) The place of George M. Beard in nineteenth-century psychiatry. *Bulletin of the History of Medicine*, 36, 245–259.

Schafer, W. (2000) *Stress Management for Wellness*. Fort Worth: Harcourt.

Selye, H. (1955) Stress and disease. *Science*, 122(3171), 625–631.

Selye, H. (1964) *From Dream to Discovery*, New York: McGraw-Hill.

Selye, H. (1976a) Forty years of stress research: Principal remaining problems and misconceptions. *Canadian Medical Association Journal*, 115(1), 53–56.

Selye, H. (1976b) Stress and physical activity. *McGill Journal of Education*, 11(001), 3–14.

Selye, H. (1976c) *Stress without Distress*. London: Transworld.

Selye, H. (1978) *The Stress of Life* (revised ed.). New York: McGraw-Hill.

Sherry, P. (1991) Person–environment fit and accident prediction. *Journal of Business and Psychology*, 5(3), 411–416.

Taylor, S.E., Klein, L.C., Lewis, B.P., Gruenewald, T.L., Gurung, R.A. and Updegraff, J.A. (2000) Biobehavioural responses to stress in females: Tend-and-befriend, not fight-or-flight. *Psychological Review*, 107(3), 411.

Teigen, K.H. (1994) Yerkes–Dodson: A law for all seasons. *Theory and Psychology*, 4(4), 525–547.

Tennant, C. (1994) Life-event stress and psychiatric illness. *Current Opinion in Psychiatry*, 7(2), 207–212.

Tyhurst, J.S. (1951) Individual reactions to community disaster: The natural history of psychiatric phenomena. *American Journal of Psychiatry*, 107(10), 764–769.

Tyhurst, J.S. (1957a) Psychological and social aspects of civilian disaster. *Canadian Medical Association Journal*, 76(5), 385–393.

Tyhurst, J.S. (1957b) The role of transition states – including disaster in mental illness. In *Symposium on Preventive and Social Psychiatry*. Washington, DC: Walter Reed Army Institute of Research and the National Research Council.

Van Harrison, R. (1978) Person–environment fit and job stress. In C.L. Cooper and R. Payne (eds), *Stress at Work*. New York: Wiley.

Van Harrison, R. (1985) The person–environment fit model and the study of job stress. In T.A. Beehr and R.S. Bhagat (eds), *Human Stress and Cognition in Organizations*. New York: Wiley.

Vingerhoets, A.J.J.M. and Perski, A. (2000) *The psychobiology of stress*. Retrieved from http://arno.uvt.nl/show.cgi?fid=12713

Wacker, J.G. (1998) A definition of theory: Research guidelines for different theory-building research methods in operations management. *Journal of Operations Management*, 16(4), 361–385.

Weber, J.G. (2010) *Individual and Family Stress and Crises*. California: Sage.

Yerkes, R.M. and Dodson, J.D. (1908) The relation of strength of stimulus to rapidity of habit formation. *Journal of Comparative Neurology and Psychology*, 18(5), 459–482.

Stress

An extensive literature in the discipline of construction management focuses on enhancing the performance of construction projects at the organisational, industrial and national levels (Fayek et al. 2006; Wong, Cheung and Fan 2009). However, there has been little or no focus on how to enhance performance at the *personal* level, even though people are the main source of success of every construction project. Among the various factors affecting individual performance, stress can be one of the most detrimental (Leung, Chan and Olomolaiye 2008). An individual suffering from stress, either physical or emotional, may lack energy, concentration and focus, resulting in a deterioration in performance as evidenced by mistakes, poor communication, absenteeism and so on. In fact, a total of 10.4 million working days was lost due to stress in the UK in 2011–2012 (Health and Safety Executive 2012). Work-related stress also incurs a monetary loss of $42 billion per annum in the US (Cousins et al. 2004). Thus, stress is not only a matter for the individual, but is also a real cost for any project, organisation, industry and even nation.

The construction industry has long been recognised as a stressful one. Stress is well known to be prevalent among its personnel, including professionals (such as project managers and cost engineers) and construction workers (Chan, Leung and Yu, 2012; Haynes and Love 2004; Leung 2004; Leung et al., 2007, 2008c; 2010, 2012). In fact, the majority of construction personnel (68%) have suffered from stress, anxiety or depression directly due to working in the industry (CIOB 2006). Driven by the explosive pace of new technology development, rapid economic growth and the increase in

Stress Management in the Construction Industry, First Edition.
Mei-yung Leung, Isabelle Yee Shan Chan and Cary L. Cooper.
© 2015 John Wiley & Sons, Ltd. Published 2015 by John Wiley & Sons, Ltd.

the number of stakeholders now involved in projects, 58% of staff agreed with the statement that the construction industry is becoming increasingly stressful (CIOB 2006). These studies indicate a close relationship between stress and construction personnel in this dynamic industry. However, it is surprising that stress management for industry personnel, such as professionals and workers, is still so uncommon. Before investigating how the stress of construction personnel can be optimised to enhance performance, it is essential to explore what stress is and how it manifests itself in an individual.

When talking about stress, people usually refer to it in terms of negative emotions or affect. However, this is only one aspect of the problem. To allow a comprehensive exploration of the manifestation of stress in an individual and in construction personnel specifically, three of the commonly delineated kinds of stress are discussed below: work, physical and emotional stress.

3.1.1   Work Stress

In terms of organisational management, stress is often referred to as the job demands made on an individual when he/she feels unable to cope adequately (Cox 1993). Developed by Lewin (1936) and Murray (1938), the *person–environment (P–E) fit* theory has long been recognised as the cornerstone of the concept of work stress in organisational research (Edwards 1996). It places equal emphasis on two variables; the individual and the environment in which he/she operates.

> It [the theory] assumes that for each individual there are environments which more or less match the characteristics of his personality. A 'match' or 'best-fit' of individual to environment is viewed as expressing itself in high performance, satisfaction and little *stress* in the system whereas a 'lack of fit' is viewed as resulting in decreased performance, dissatisfaction and *stress* in the system. (Pervin 1968: 56)

The P–E fit theory addresses the fit or interaction between an individual and his/her work and general living environment. In the organisational context, this concept mainly focuses on the fit or lack of it between the job demands made of an individual (such as the expectations of supervisors or others) and his/her actual abilities.

3.1.1.1 *P–E fit between expected and actual ability*

Stress arises if there is a misfit between the person and the environment (French and Caplan 1972; French et al. 1974; Harrison 1978). The P–E fit refers to the gap between the perceptions of an individual of the fit between the job environment, in terms of demands made and his/her abilities or characteristics as expressed within that environment. Based on this concept, research has focused on the fit between supplies (the resources available in the environment) and values (the preferences and wishes of the person) (Kristof-Brown, Zimmerman and Johnson 2005). The causality of the P–E fit concept is illustrated in Figure 3.1.

Construction personnel (including both professionals and technical workers) are generally well trained and assessed in terms of their technical education and its practical aspects, but individual perceptions of ability and value will be different from one person to another. A P–E misfit does not necessarily lie in the environmental variables relevant to an individual, such as job demands and supplies, but may also stem from individual ability, preferences and values. There are generally three potential relationships between the P–E fit dimensions and individual stress levels (French et al. 1974): monotonic, asymptotic and/or U-shaped (French et al. 1982; Harrison 1978) (see Figure 3.2). A *monotonic* relationship indicates that the stress level of an individual increases along with the demand from the environment. An *asymptotic* relationship means that an individual has spare capacity in terms of ability, so demand does not influence his/her stress level, which remains constant in the demand < ability phase. The equilibrium point of P–E fit (i.e., where demand = ability) represents the optimisation of stress levels. A U-shaped relationship implies the presence of excessive supplies for one stressful stimulus, resulting in deficient supplies for another (French et al. 1974).

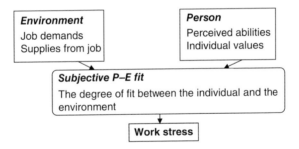

Figure 3.1   How work stress results from P–E fit.

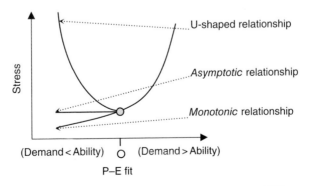

Figure 3.2   Three shapes of the P–E fit curve (adopted from Harrison 1978).

It follows from this that work stress results from the individual's cognitive evaluation of his/her P–E fit. This in turn depends on perceptions of his/her ability to fulfil the task in the context of the anticipated difficulty or challenge involved in the job of an individual (Gmelch 1982; Leung 2004; Leung et al., 2008a, 2011). This is different from work overload, which is a well-recognised stressor (i.e., a source of stress; refer to Chapter 4 for more details) concerning job demands from the external environment only. Work stress is induced by the deviation between the expected ability of an individual (i.e., the environment) and his/her actual ability (i.e., the personal) to deal effectively with tasks (Monat and Lazarus 1991).

3.1.1.2 *Expectancy and valence in work stress*

P–E fit is a subjective perception which differs from individual to individual. When placed in the same external environment, different people can form different perceptions about it and draw different levels of satisfaction from working within it. The question then arises: what determines the extent to which the ability of an individual and his/her personal values and the role played by the fit or misfit between that individual and the environment, each contribute to the stress experienced? Vroom's valence-expectancy theory further elucidates these individual differences in subjective perceptions of P–E fit.

The valence-expectancy theory is a theory of motivation which focuses on conscious and psychological involvement of individuals in choosing between various actions (Vroom 1964). It is one of the dominant process theories adopted in motivation research. Individuals have expectations and/or anticipation of the outcomes of their behaviour (Lewin 1936; Tolman 1952) and are motivated by the perceived probability and incentive value of success (Atkinson 1957). In the motivation process, expectancy and valence represent the essential cognitions of individuals in making choices about their behaviours (Vroom 1964). Expectancy refers to individual beliefs or expectations about whether a certain course of actions or behaviour will result in particular outcomes (Nadler and Lawler 1983; Pinder 1991; Tolman 1952). Vroom (1964) defines the term valence as the affective orientation people hold with regard to such outcomes. Other researchers refer to this as anticipated satisfaction (Diefendorff and Croyle 2008; Lee 2007), experience satisfaction (Chiang and Jang, 2008; Wetzels et al., 2010) and importance (Kominis and Emmanuel 2006; Nadler and Lawler 1983).

Vroom's valence-expectancy theory proposes that if an individual, after assessing the effort (E) that he/she needs to make in order to result in a certain performance (P) (E→P expectancy), anticipates that performance (P) will lead to a certain set of outcomes (O) (P→O expectancy) and predicts that the outcome will be valuable and rewarding (valence), then that individual will be motivated to invest effort in the task. Valence, positive or negative, is closely related to affect and to the pleasant or unpleasant feelings experienced by an individual which directly influence his/her readiness to take an action or make an effort (Russell 2003). It is the pleasurable affect resulting from the favourable appraisal of an individual of his/her performance at work which facilitates him/her to achieve what is valued on the job by making an effort to deliver (Locke 1969). If an individual anticipates satisfaction from a job outcome (i.e., a positive valence), he/she will be more likely to perceive a better fit between the environment and his/her personal needs. Such a perception is more likely to create high expectations of the effort required to achieve a predicted level of performance. Such positive valence, together with high expectations, not only directly reduces work stress, but also do so indirectly by motivating the individual to make an effort to reduce misfit. Therefore, valence and expectancy are positively associated with positive affect (Erez and Isen 2002).

On the other hand, the concept of coping behaviours is congruent with the notion of effort in Vroom's theory, since it is concerned with effort in response to a broader situation, including a stressful one. Hence, based on this theory, one can conceptualise the relationships between individual motivation to cope, coping behaviours, stress and performance as shown in Figure 3.3, which captures the roles of expectancy and valence in stress management. An individual will be motivated to cope with stress by undertaking certain coping behaviours if he/she has high expectations that these behaviours will be influential (i.e., CP→S expectancy), that the reduction of stress will improve performance (i.e., S→P expectancy) and that the resulting satisfaction with such performance will be high (valence).

In the construction context, a lot of P–E fit factors can cause work stress, such as the number of project deadlines (particularly true for senior professionals who need to handle multiple projects at the same time and for technical workers who are often forced to work overtime so as to complete physically demanding tasks within a tight timescale); the number of tasks (such as numerous meetings and phone calls, frequent documentation submissions, site visits); difficulty levels of jobs (such as limited funding, making difficult decisions, multitasking programmes, resource allocation); degree of complexity of the work and so on (see e.g. Leung et al., 2008b, 2008c, 2012). It is not solely these environmental factors which determine the level of work stress of a construction personnel, but also his/her perceptions

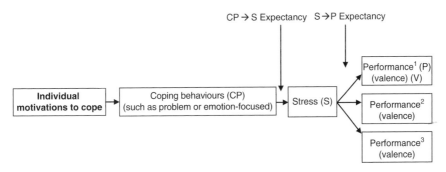

Figure 3.3 Motivation to cope, coping behaviours, stress and performance in the expectancy theory. Note: Motivation to cope equals [CP → S] x Σ[(S → P)(V)].

of the fit between these job demands and his/her ability to deal with them. These, in turn, are affected by the valence and expectations of the task. People who lack the ability to multitask across various projects with limited resources and multiple stakeholders will find it difficult to adapt to the working environment of the construction industry. On the other hand, personnel who expect to face challenges in complicated construction projects and have the ability to do so, but are assigned only repetitive tasks such as measurement, copying and so on, may also feel boredom and the resulting stress.

3.1.2   Physical Stress

The roles of most construction personnel require not only cognitive and creative abilities, but also the physical capacity to supervise or execute various tasks on site in changeable weather conditions (rain or shine, hot or cold). Such roles make higher demands on their physical capabilities compared to other industries, so an understanding of physical stress is also essential to the work performance of construction personnel. The physical aspect of stress can be defined as a state of threatened homeostasis resulting from exposure to stressors (Chrousos and Gold 1992; Teasdale and McKeown 1994). Developed by Cannon (1939), the concept of homeostasis refers to a complex dynamic equilibrium state within human beings, which is often stimulated by various external and internal stressors. When an individual is exposed to a stressful situation, homeostasis is influenced and then re-established by a complex adaptive stress–response mechanism which has both behavioural and physiological aspects.

In stressful circumstances, hormones are released from the brain in order to help other parts of the body to cope with the changes. The body then tries to overcome stress by making various physical adjustments. After a certain period of time, these adjustments can revert back to normal if the body is no longer affected by any stressor. This process, which aims to maximise adaptation for survival, prepares the body for the well-known 'fight or flight' response (see Section 2.2 for more details). However, if stressful conditions are continuously present, these physiological reactions may persist, become maladaptive and result in physical stress symptoms which are harmful to the individual metabolism, immune system, growth and so on (McEwen 1998; Teasdale and McKeown 1994).

3.1.2.1   Physiological reactions

Construction personnel maintain a complex and dynamic state of homeostasis which is constantly stimulated by external factors. In stressful situations, the adaptive homeostatic system activates compensatory responses so as to overcome these stressors. This leads to adjustments in two main physiological systems: the central nervous system, constituting the *hypothalamic–pituitary–adrenal (HPA) axis* (see Figure 3.4) and the periphery, constituting the *sympathetic nervous system (SNS)* (Chrousos 2009; Chrousos and Gold 1992; Gilbey and Spyer 1993; Kventnansky, Sabban and Palkovits 2009).

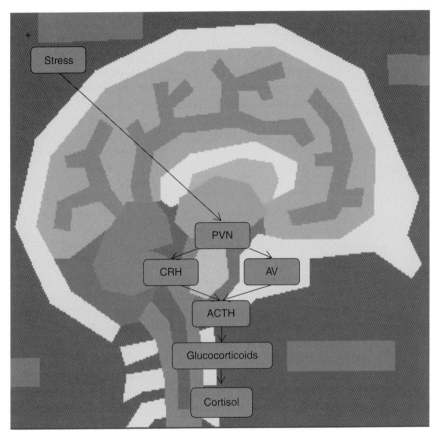

Figure 3.4 Hypothalamic–pituitary–adrenal (hpa) axis. Note: PVN (paraventricular nuclei); CRH (corticotrophin-releasing hormone); AV (arginine-vasopressin); ACTH (adrenocorticotropic hormone).

The *central nervous system is strategically located in the hypothalamus and the brain stem* and is comprised primarily of the following mediators: corticotrophin-releasing hormone (CRH), arginine-vasopressin (AV), the paraventricular nuclei (PVN) of the hypothalamus, the pro-opiomelanocortin-derived peptides α-melanocyte-stimulating hormone and β-endorphin. By secreting CRH and AV into the portal system, the hypothalamus regulates the level of adrenocorticotropic hormone (ACTH), which further stimulates the adrenal cortex to secrete glucocorticoids. Glucocorticoids influence cortisol secretion. They provide feedback at the hypothalamic and pituitary levels and to the higher brain centres and so are essential in controlling the basal activity of the HPA axis and various stress responses. Through these mediators and their effects, the central nervous system reacts to stress by facilitating and inhibiting different neural pathways. It facilitates the neural pathways for acute, nonpermanent adaptive functions such as arousal, alertness, attention and aggression, while inhibiting those for vegetative, nonadaptive functions such as feeding, growth and reproduction.

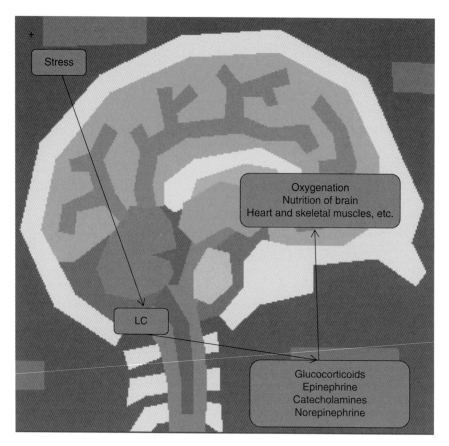

Figure 3.5   Sympathetic nervous system (SNS). Note: LC (locus ceruleus).

In the *sympathetic nervous system (SNS)*, the innervations of the peripheral organs emanate from the efferent preganglionic fibres, whose cell bodies
are located in the intermediolateral column of the spinal cord (see Figure 3.5).
Through the mediators of glucocorticoids, epinephrine, catecholamines,
norepinephrine from the adrenal medulla and so on, the peripheral organs
and tissues react to stress by, for example, increasing oxygenation; increasing
nutrition to the brain, heart and skeletal muscles; increasing cardiovascular
tone and respiration; speeding up the metabolism; increasing the detoxifica-
tion of the metabolism and the gastrointestinal functions.

However, it is not the case that the more vigorous the activity of the homeostatic system, the better the effects. According to Chrousos and Gold (1992),
the relationship between homeostatic system activity and its effects is an
inverted U-shape, not linear (see Figure 3.6). This implies that neither an excessive nor deficient level of homeostatic system activity will yield optimum
effects. Only a medium level can result in adaptive homeostasis, returning to
basal homeostasis, a state termed *eustasis* (centre of the curve). Both hyper- and
hypofunction of the homeostatic system can result in defective homeostasis
and distress, a state labelled *allostasis* (either side of the curve). This is harmful

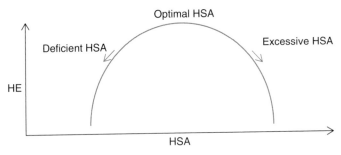

Figure 3.6 Inverted U-shaped relationship of the homeostatic system and its effect. Note: HE (homeostatic effect); HSA (homeostatic system activity).

to the organism in terms of both morbidity and mortality. There is, furthermore, a third state, in which optimisation is achieved on a given occasion and the organism then gains from this experience to adjust the homeostatic capacity for later stimulation. Chrousos (2009) describes this as hyperstasis.

3.1.2.2 *Physiological stress symptoms*

On the other hand, self-restrained adaptive stress responses are important to successful outcomes (Kyrou and Tsigos 2009). The adaptive homeostatic system, mainly the HPA axis and the SNS, does not aim to make permanent changes to the body. It is supposed to revert back to the normal state after a certain period of time, once the individual has fought off or escaped from the trigger(s). In the process of restoring such balance, the associated temporary respiratory, gastrointestinal, anti-growth, anti-reproductive and immunosuppressive effects cease in a process which facilitates normal bodily survival (Chrousos 2009; Kyrou and Tsigos 2009). However, chronic or escalating allostasis can result in allostatic overload (i.e., the wear and tear that results from either too much stress or from inefficient management of allostasis), which in turn leads to a prolonged stress response and pathophysiology (McEwen 2007). In other words, chronic activation of the adaptive stress response system prolongs the associated effects, causing pathological and adverse physical conditions in affected individuals.

Stress leads to various physiological symptoms due to the secretion of the various mediators (or hormones) in the adaptive stress response process, such as CRH, cortisol, glucocorticoids and norepinephrine. Some forms of stress have only a short-term effect, while others can have longer term implications. Short-term stress may occur during difficult meetings, sporting events or confrontational situations, but the effects may not be significant if the stimulus passes quickly. In contrast, however, long-term stress can induce physical problems and psychological fatigue, which affect health and undermine team morale (see Figure 3.6) (Bowles and Cooper 2009). For instance, the immune-CRH-induced degranulation of mast cells in vulnerable organs (such as the skin or lungs) or in meningeal blood vessels could cause allergies such as eczema or asthma, or headaches due to vasodilatation and increased permeability of the blood–brain barrier, respectively (Chrousos 2009). On the other hand, CRH, cortisol and norepinephrine also affect the sleep system, causing disorders such as insomnia and poor sleep quality. In general, the mediators of the adaptive stress response process may lead to allergies (such as eczema, asthma and dermatitis); pain (such as headaches and back, abdominal and musculoskeletal pain); and gastrointestinal disorders (such as loss of appetite, constipation and diarrhoea) among other effects (Chrousos and Kino 2007; McEwen 2007; Montoro et al. 2009). Research has been carried out to investigate such physical stress symptoms in construction personnel, including headaches, appetite loss, sleep disorders, skin problems, gastrointestinal disorders and musculoskeletal pain (Leung et al. 2011, 2012).

3.1.3   Emotional Stress

Hormones or mediators like cortisol, as byproducts of the physiological stress response, are also associated with emotional states. For instances, levels of the HPA-axis hormone cortisol have been found in various studies to be correlated to some extent with measures of burnout (Mommersteeg et al. 2006; Pruessner, Dirk and Clemens 1999). A high level of CRH is associated with anxiety and depression (Holsboer 1999; Reul and Holsboer 2002). It can be seen that there is a close relationship between the physical and emotional stress responses of the individual. However, there is still no certainty as to whether emotional stress causes a physiological response, or the physiological response affects the emotional state of an individual. It is nevertheless certain that stress can manifest itself in individuals in the form of emotional states such as anxiety, depression, displeasure and nervousness (Daniels and Guppy 1994; Nyssen et al. 2003).

3.1.3.1   *Psychological stress reactions and emotions*

Psychological stress reactions are the negative cognitive and emotional states triggered by stressful factors in the work or living environment of an individual (Cohen, Tyrrell and Smith 1991; Lazarus and Folkman, 1984). These reactions are generated by our brains. Cognitive processes, such as thinking, reasoning, memory and problem solving, are governed by the neocortex, while emotion is regulated by the *limbic system*. This is made up of the limbic cortex and the structures of the brain stem with which it has primary connections, including the rhinencephalon (MacLean, 1949, 1952, 1972). Although the exact brain areas covered by the limbic system remain undefined, the role of the limbic system in emotion is supported by various experimental data collected from tests of animals such as monkeys and rats. These results indicate that the limbic system obtains information in terms of 'feelings that promote self-preservation and preservation of the species' in the process of feeding, fighting and self-protection (MacLean 1994: 108). The limbic system is also closely related to various emotional behaviours displayed by human beings, such as crying on separation from the nursing mother and laughing while engaged in play.

Before exploring the nature of the limbic system, it is also important to discuss the nature of emotion itself. *Emotion* refers to a natural instinctive state of mind deriving from circumstances, mood or relationships with others (Oxford Dictionary 2012). As distinct from sensation, emotion refers not only to feelings, like anger or interest, but their combination with thoughts (Wierzbicka 2010): 'emotion words are not names of things – rather they demarcate mental representations that are constituted as feelings of pleasure or displeasure and socially situated conceptualizations of emotions' (Barrett et al. 2007: 390). In other words, the term emotion is culture- and language-specific, fuzzy, relates to family resemblance and is an ever-evolving cluster (Russell 1991; Wierzbicka 2010). In addition, both descriptive and prescriptive definitions of emotion are used in everyday life and the process of developing scientific theories, respectively (Widen and Russell 2010).

According to Descartes (1967), emotion can be defined as a behavioural manifestation identified with a particular subjective affect in an individual. This definition is used in MacLean's explanation of the limbic system (1994). When the neocortex receives information from the external environment (such as stimulation or stressors), the limbic cortex also receives a large amount of information from the internal environment, meaning the person's previous experiences of the situation which contribute to his/her subjective affect. This is called the limbic effect, where the thoughts, propensities and subjective affect of an individual trigger each other reciprocally. In fact, this concept aligns with that of P–E fit discussed earlier (see Section 3.1.1), within which stress results from the process of appraising the fit between the environment (external world) and person (internal world). Emotional stress mainly focuses on the products of the automatic limbic-generation process. Although our information about how emotions are generated in the brain is still limited, it is clear that the limbic system has a close association with one's emotional states.

Exploring the fear system may help us understand more deeply the mechanism of emotion, or more specifically the conditioned *emotional response*, experienced by our bodies (LeDoux 2000). Instead of taking into account only the subjective experience of an individual, the fear system covers the process through which one detects and responds to danger. The fear system was developed based on classical conditioning theory, which proposes that there are two types of stimulus, conditioned and unconditioned. A conditioned stimulus is an initially neutral property, such as light or sound. An unconditioned stimulus, which follows the conditioned one, is some form of biologically significant stimulus, such as pain or food. There are many different types of conditioning procedures, but in all of them, a conditioned stimulus generally signals the subsequent occurrence of an unconditioned one. Information about conditioned and unconditioned stimuli is firstly transmitted to the amygdala, which then sends a message to the brain stem to control the fear reactions of an individual in terms of a behavioural, autonomic and endocrine response. In fact, the emotional response system serves to protect an individual from harm (Carlson 2010). For instance, when the conditioned stimulus is being on a high platform and the unconditioned stimulus is the pain of falling, the result may be a conditioned fear of heights. A construction worker in this position may then be more cautious about observing safety measures for working at height. However, if such reactions persist or become inappropriately restricting, such a mechanism would become maladaptive.

3.1.3.2 *Emotional stress symptoms*

Emotional stress occurs when individuals become emotionally drained and chronically fatigued and lose the ability to accomplish tasks due to prolonged exposure to stressful conditions in their work, personal life or relationships (Freudenberger 1983). The symptoms of emotional stress for construction personnel can include fatigue, a feeling of being 'used up', worrying about work and being emotionally drained as a result of work (Leung et al., 2010). Prolonged emotional stress can result in emotional exhaustion (Cordes and Dougherty 1993), which occurs when individuals become chronically fatigued and lose the ability to give adequate attention to their work. According to Cordes and Dougherty (1993), such exhaustion may be combined with feelings of frustration and tension. Other symptoms of emotional exhaustion include disillusionment with the job, a loss of meaning, cynicism towards organisations or clients, feelings of helplessness and frustration, a sense of lacking the power to change events, strong feelings of anger with the people felt to be responsible for the situation and feelings of depression and isolation.

In fact, *emotional exhaustion* is one of the three widely adopted burnout symptoms as defined by Maslach (1996). Burnout, defined as a state of physical, emotional and mental exhaustion caused by long-term involvement in emotionally demanding situations (Aronson and Kafry 1998), has long been recognised to have a significant impact on construction personnel (Leung et al., 2008a; Yip and Rowlinson 2009). As well as emotional exhaustion, burnout encompasses two other symptoms, namely depersonalisation and reduced personal accomplishment (Janssen et al. 1999; Maslach et al. 1996; Schaufeli et al. 1993). Depersonalisation is considered to be a negative response to emotional exhaustion. Common symptoms include physical withdrawal, poor relationships with others and making numerous complaints about the workplace (Cordes and Dougherty 1993). For example, construction personnel may complain about their job or worksite and avoid communicating with colleagues or friends. Reduced personal accomplishment denotes a tendency to engage in negative self-evaluation and to have poor self-esteem, dissatisfaction with work performance and poor relationships. Individuals suffering from this may feel not only that they are not making progress, but indeed that they are "losing ground" (in other words, they have low motivation and commitment at work due to their prolonged frustration) (Cordes and Dougherty 1993).

However, *emotional exhaustion* is the main focus in our studies of construction personnel, for two reasons. Firstly, it has been identified as the most critical dimension of burnout, while the other two are merely theoretically related variables which address different concepts (Shirom and Ezrachi 2003). Secondly, the concepts of depersonalisation and reduced personal accomplishment may have some similarities with workgroup relationship stressors and work stress, which are discussed elsewhere in this book. In fact, the emotional state of the individual has been found to influence his/her physiological states, resulting in symptoms such as migraine, insomnia, impaired immune response affecting susceptibility to disease and the exacerbation of pre-existing medical disorders (see e.g. Carroll and White 1982; Cohen, Tyrrell and Smith 1991; Hung, Liu and Wang 2008; Leung, Chan and Olomolaiye 2008; Pines 1982). Moreover, there is a very close relationship between psychological and physical stress. Emotional stress is also significantly associated with the physical stress of construction project managers (Leung, Chan and Chen 2010).

3.2   Development of a Conceptual Model of Stress

In summary, stress can be categorised into three major domains: work, physical and emotional. The mismatch between a person's actual ability and the external demands of work elicits work stress, which can manifest itself physically or emotionally. In addition, the physical and emotional stress states of an individual are also closely linked. In other words, the three kinds of stress are not separate mechanisms but influence the individual through their close interrelationships. The overall stress model is illustrated in Figure 3.7.

A number of studies have looked at the interrelationships between the three kinds of stress for construction estimators, project managers and other professionals as well as construction workers (Leung, Chan and Olomolaiye 2008; Leung, Chan and Chen 2010; Leung, Chan and Yu, 2012). Professional estimators and other estimating personnel can be considered as having the same levels of work and emotional stress in the design stage of projects (Leung, Chan and Olomolaiye 2008). Professional estimators experience more work than emotional stress. Construction personnel often undertake tasks involving calculation, planning and organisation, which are mainly of an objective nature. They should not wait for emotional stress to occur before taking appropriate action. Moreover, emotional and physical stress are also interrelated for construction workers, being triggered by poor on-site working environments.

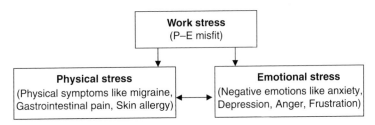

Figure 3.7 Conceptual model of stress (adopted from Leung, Chan and Olomolaiye 2008; Leung, Chan and Chen 2010).

A previous study of construction project managers show that work stress is the antecedent of emotional stress, which itself can then predict physical stress (Leung et al. 2010). It is interesting to note that work stress directly affects emotional rather than physical stress for project managers. This may be because the physical conditions they face are tougher, since they are often required to work on site and deal with adverse environments. Their ability to respond to this stressor may be stronger than their level of emotional resilience. If so, this would mean that their physical condition was less susceptible to work stress. This may also explain why physical stress has no impact on emotional stress for this group, whereas in reverse, emotional stress is shown to trigger stress-induced physical problems. Stress among construction project managers manifests itself not only psychologically, but physically. The individual may suffer from headaches, back pain, loss of appetite or skin problems due to the harsh and demanding working environment in the office or on site. In general, if project managers fail to cope effectively with these high levels of work stress, they are likely to suffer from physical and emotional stress as well.

3.3   Research Results on Stress among Construction Personnel

This section discusses evidence obtained from a scientific investigation of how stress manifests itself in construction personnel as a specific group. The research has both statistical and case study components, both of which are reported here and is based on an extensive literature developed over three decades of research on the three types of stress.

3.3.1   Statistical Studies

This study used questionnaires to explore the stressors experienced by professionals and manual labourers in the construction industry. Surveys were designed and distributed to 2000 construction employees who work in different positions and disciplines (e.g., project managers, surveyors, engineers, architects and construction workers) and in different types of construction companies (e.g., developer, public sector, consultant firm, main contractor and subcontractor). The surveys were administered by email, fax and post with a prior verbal telephone conversation with all targeted construction companies – 702 valid surveys were received, representing a response rate of 35.1%. Data analyses were conducted using SPSS 20.0. Among the 702 participants, 39.3% were construction workers, 32.5% project managers, 10.8% architects, 6.8% engineers, 5.8% quantity surveyors and 4.3% others.

3.3.1.1 *Stress of construction personnel*

When they talk about 'stress', people generally mean overstress and its effects. Though too much stress (i.e., overstress) can indeed result in burnout, too *little* stress (understress) can also affect performance through rustout. To measure the work stress levels of construction personnel, the rustout–burnout (RO–BO) scale developed by Gmelch (1982) was adopted. This measures work stress on the basis of the difference between the actual and expected ability of an individual to achieve tasks normally assigned to construction personnel. Such an indirect measurement minimises the effects of subjective assessment and thus has been widely adopted in stress management studies in the field of construction (Leung, Chan and Olomolaiye 2008; Leung, Skitmore and Chan, 2007; Leung, Chan and Chen, 2011). To ascertain these differences, respondents were asked to rate both their (A) expected and (B) actual abilities using a Likert-type scale ranging from 1 (none) to 7 (a great deal) in terms of various job-related dimensions (see Table 3.1). Their work stress was then measured by calculating the differences between (A) and (B) across the various items, ranging from –6 to +6. A deviation within the range –1.4 to +1.4 is considered optimal using the current 7-point Likert-type scale form of the survey (range calculated based on the optimisation zone developed by Gmelch (1982), who adopted a 5-point scale). The results of the survey showed that the mean scores of all the work stress items lay within the optimisation zone. This means that the construction personnel taking part in this study considered their actual ability to deal with the difficulty and complexity of their work to be slightly higher than their expected ability (–0.120 and –0.219, respectively), while their expected ability across all the other items was higher than their actual skills (from +0.022 to +0.255). Among all seven items, the highest positive deviation between expected and actual ability was found for the number of projects (+0.255).

To measure the physical stress levels of construction personnel, eight items were selected and adopted from previous studies (Leung, Chan and Olomolaiye 2008; Leung, Chan and Chen 2010; Li Calzi et al., 2006). The respondents were asked to rate the extent to which they suffered from various physical symptoms (listed in Table 3.2) using a 7-point Likert-type scale

Table 3.1   Work stress of construction personnel (N = 702).

Work stress items	Expected ability (A)	Actual ability (B)	Mean dev. (A – B)	Std dev.
1. The number of projects	**I have to handle**	**I am capable of handling**	**0.255**	**1.236**
2. The number of tasks	I have to do	I am capable of doing	0.022	1.221
3. The degree of complexity of work	I have to do	I am capable of doing	–0.219	1.175
4. The number of project deadlines	I have to meet	I am capable of meeting	0.014	1.107
5. The responsibility of my work	I have to take	I am capable of handling	0.013	1.213
6. The level of difficulty in my work	I have to deal with	I am capable of dealing with	–0.120	1.205
7. The quality of work	I have to do	I am capable of doing	0.179	1.180

Note: **Bolded** item - item with the highest deviation between expected and actual aibility obtained.

Table 3.2   Physical stress of construction personnel.

Physical stress symptoms	Sample Size	Mean	Standard dev.
1. Headaches and migraines	702	3.110	1.667
2. Losing appetite	**702**	**3.229**	**1.560**
3. Sleep disorders	205	3.117	1.601
4. Skin disorders	562	2.931	1.717
5. Gastrointestinal disorders	136	3.074	1.604
6. Musculoskeletal pain	**425**	**3.657**	**1.686**
7. High blood pressure	402	2.816	1.777
8. Sweating, palpitation, trembling	**425**	**3.233**	**1.663**

Note: **Bolded** items - items with the highest mean score.

Table 3.3   Emotional stress of construction personnel.

Emotional stress symptoms	Sample Size	Mean	Standard dev.
1. Worry	**562**	**3.940**	**1.556**
2. Fatigue	**484**	**3.957**	**1.532**
3. Used up	**484**	**4.310**	**1.520**
4. Emotionally drained	484	3.831	1.562
5. Frustrated	194	3.742	1.417
6. Unhappy	135	3.652	1.289
7. Losing temper	126	3.619	1.317
8. Anxious	**136**	**3.963**	**1.330**

Note: **Bolded** items - items with the highest mean score.

ranging from 1 (rarely true) to 7 (usually true). The results indicated that the mean scores of all the physical stress items were below the neutral point of 4, while the highest scores were found for loss of appetite; sweating, palpitations and trembling; and musculoskeletal pain.

Eight items measuring emotional stress symptoms were also used in this survey (Leung, Chan and Olomolaiye 2008; Leung, Chan and Chen 2010). Respondents were asked to rate the extent to which they suffered from each symptom using a 7-point Likert-type scale ranging from 1 (rarely true) to 7 (usually true). The results, as listed in Table 3.3, showed that participants scored highest on feeling used up at the end of the working day, being worried about their job, fatigue and getting anxious.

Due to the physically demanding nature of the jobs done by many construction personnel and the crisis-ridden site environment, health care promotion in this sector has long been dominated by training in physical well-being. However, the Tables 3.2.and 3.3 indicate that construction personnel generally experienced more emotional than physical stress symptoms. This sheds light on the importance of having construction stakeholders monitor and promote not only the physical well-being of their staff, but also their emotional health through means such as offering mental health first aid courses and the type of mindfulness-based stress management workshops which are now common in various institutions.

3.3.1.2 *Stress of different construction occupational groups*

In order to investigate the vulnerability of different types of construction personnel to the three main kinds of stress, the unit of analysis in this section changes from item to factor. The work stress levels of construction personnel were calculated by summing the differences between (A) and (B) obtained from the seven items listed in Table 3.1, while their physical and emotional stress levels were computed by summing the ratings of the eight items listed in each of Tables 3.2 and 3.3, respectively. According to the reliability analysis, the alpha value of this measure of physical stress was 0.771 and of emotional stress 0.733, representing high internal consistency between these two factors (Hair et al. 1998). As shown in Table 3.4, a one-way between-groups analysis of variance (ANOVA) was conducted to explore the three types of stress suffered by various groups of construction personnel. Subjects were divided into six groups according to their professional discipline, including (1) project managers (including construction project managers, contract managers and directors); (2) architects; (3) engineers (including structural, building services and general building engineers); (4) quantity surveyors; (5) construction workers; and (6) others (such as building surveyors, safety officers, site agents, foreman, facilities managers). However, there were no significant differences at the $p < 0.05$ level in the work and emotional stress scores across all five groups ($F = 1.220$ and $F = 2.024$; sig > 0.05). This indicates that the prevalence of work and emotional stress was similar for all survey participants. However, there was a significant difference in levels of physical stress between the groups ($F = 2.518$, sig < 0.05; see Figure 3.8).

To identify more precisely the source of this significant difference in levels of physical stress, a *post hoc* comparison was conducted using the Tukey HSD test. The results, shown in Table 3.4, indicate the mean score of Group 1 (project managers) was significantly lower than that of Groups 4 (quantity surveyors) and 5 (construction workers), with mean differences of –2.396 and –1.042, respectively. Compared with project managers, construction workers have to make much more physical effort to carry out tasks on site, such as delivering materials, controlling heavy equipment and climbing up a hoist. These physically demanding tasks undoubtedly affect their bodily well-being and induce physical stress symptoms such as back pain and sweating. However, it is interesting to note that the quantity surveyors scored highest on physical stress among the six groups. When compared to other personnel, quantity surveyors tend to spend more of their working day in the office rather than on site. It may be speculated that the substantial amount of paperwork involved in the (often urgent) estimation tasks carried out by surveyors may be responsible for them reporting more physical stress symptoms (such as appetite loss) than the other groups (Table 3.5). Many quantity surveyors also work in poorly designed office spaces with inappropriate posture and take inadequate amounts of physical exercise.

Table 3.4 One-way between-groups ANOVA for stress levels of construction personnel.

Construction personnel groups	N	Mean	SD	F (ANOVA)	Sig. (ANOVA)	Sig. (Levene)	Group	Mean Diff.	SD	Sig.
Work stress										
1. Project managers	228	0.579	5.398	1.220	0.298	0.000				
2. Architects	76	−0.171	1.649							
3. Engineers	48	1.115	5.453							
4. Quantity surveyors	41	−0.427	6.912							
5. Construction workers	276	0.070	3.522							
6. Others	30	1.317	4.207							
Total	702	0.322	4.507							
Physical stress										
1. Project managers	228	23.518	6.022	**2.518**	**0.028**	**0.000**	QS	**−2.396**	**0.791**	**0.003**
							CW	−1.042	0.417	0.013
2. Architects	76	24.392	1.058							
3. Engineers	48	24.678	5.148							
4. Quantity surveyors	41	25.914	4.141							
5. Construction workers	276	24.560	3.850							
6. Others	30	24.025	5.069							
Total	702	24.266	4.688							
Emotional stress										
1. Project managers	228	31.240	3.632	2.024	0.073	0.000				
2. Architects	76	31.362	1.716							
3. Engineers	48	32.018	3.278							
4. Quantity surveyors	41	31.965	4.117							
5. Construction workers	276	31.006	2.830							
6. Others	30	32.455	5.185							
Total	702	31.309	3.275							

Note: QS - Quantity surveyors; CW-Construction workers.
Bolded figures - Significant between-group differences

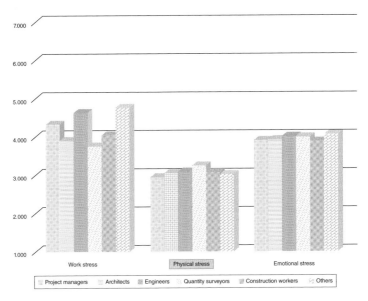

Figure 3.8   Stress scores for construction personnel in different roles. Note: The −6 to 6 scale of work stress is converted to 1–7 for the sake of comparison with physical stress and emotional stress. Highlighted items – Significant differences in one-way ANOVA (refer to Table 3.4)

3.3.1.3   Stress of different age groups

Another one-way between-groups ANOVA was conducted to explore the difference in reported work, physical and emotional stress for personnel in different *age* groups (see Table 3.6). Subjects were divided into four groups: (1) aged 29 or less; (2) 30 to 39; (3) 40 to 49 and (4) 50 or above. There were no significant differences at the $p < 0.05$ level in the physical and emotional stress scores for the four age groups ($F = 1.958$ and $F = 2.350$; sig > 0.05). However, a significant difference at the $p < 0.05$ level was found in the work stress scores ($F = 2.754$: see Figure 3.9). In addition, for the work stress test, the significant value for Levene's test (0.812) was greater than 0.05, demonstrating no violation of the assumption of homogeneity of variance. Personnel aged 50 or above were the only age group to report negative mean scores for work stress, meaning that in general, their actual ability was higher than the work demands exerted by the environment. On the other hand, the younger the personnel, the more work stress they suffered (i.e., work demands exceeding their actual ability).

To identify more precisely the source of this significant difference between the age groups, a *post hoc* comparison was conducted using the Tukey HSD test. The results, shown in Table 3.6, indicate that the mean score for Group 4 (aged 50 or above) was significantly lower than that of Groups 1 (29 or below; −1.407), 2 (30–39; −1.237) and 3 (40–49; −1.304). The difference in work stress scores between Groups 1 and 4 was the largest. This is readily understandable, as construction personnel aged 50 or above are equipped with a richer set of practical experiences which they can use to solve the various problems arising at work and so will be more capable of coping with external work demands.

Table 3.5 One-way between-groups ANOVA for physical stress of construction personnel.

Construction personnel groups	N	Mean	SD	F (ANOVA)	Sig. (ANOVA)	Sig. (Levene)	Group	Mean diff.	SD	Sig.
Headaches and migraines										
1. Project managers	228	2.733	1.583	1.220	0.298	0.000				
2. Architects	76	3.079	1.598							
3. Engineers	48	2.958	1.429							
4. Quantity surveyors	41	3.610	1.672							
5. Construction workers	276	3.362	1.705							
6. Others	30	3.300	1.860							
Total	702	3.110	1.667							
Losing appetite										
1. Project managers	228	3.140	1.578	**2.876**	**0.014**	**0.017**	**QS**	**−0.967**	**0.300**	**0.017**
2. Architects	76	2.790	1.427							
3. Engineers	48	3.249	1.414							
4. Quantity surveyors	41	3.756	1.640							
5. Construction workers	276	3.033	1.650							
6. Others	30	3.323	1.550							
Total	702	3.229	1.560							
Sleep disorders										
1. Project managers	228	3.144	0.998	0.242	0.944	0.670				
2. Architects	76	3.094	1.683							
3. Engineers	48	3.073	1.012							
4. Quantity surveyors	41	3.178	0.823							
5. Construction workers	276	2.994	0.503							
6. Others	30	3.119	0.864							
Total	702	3.117	1.601							
Skin disorders										
1. Project managers	228	2.817	1.517	0.697	0.625	0.000				
2. Architects	76	2.907	1.760							
3. Engineers	48	3.123	1.250							

	N	Mean	SD			
4. Quantity surveyors	41	3.197	1.079			
5. Construction workers	276	2.845	0.469			
6. Others	30	2.967	1.668			
Total	702	2.931	1.717			
Gastrointestinal disorders						
1. Project managers	228	3.035	1.050	0.828	0.530	0.000
2. Architects	76	3.119	0.388			
3. Engineers	48	3.213	1.148			
4. Quantity surveyors	41	3.150	0.575			
5. Construction workers	276	2.959	0.494			
6. Others	30	3.073	0.704			
Total	702	3.074	1.717			
Musculoskeletal pain						
1. Project managers	228	3.498	0.966	1.108	0.355	0.000
2. Architects	76	3.793	1.586			
3. Engineers	48	3.657	0.000			
4. Quantity surveyors	41	3.657	0.000			
5. Construction workers	276	3.750	1.701			
6. Others	30	3.657	0.000			
Total	702	3.657	1.686			
High blood pressure						
1. Project managers	228	2.715	1.412	0.869	0.502	0.000
2. Architects	76	2.792	0.208			
3. Engineers	48	2.660	1.073			
4. Quantity surveyors	41	2.844	1.373			
5. Construction workers	276	2.697	1.776			
6. Others	30	2.933	1.428			
Total	702	2.816	1.777			

(Continued)

Table 3.5 (Continued)

Construction personnel groups	N	Mean	SD	F (ANOVA)	Sig. (ANOVA)	Sig. (Levene)	Group	Mean diff.	SD	Sig.
Sweating, palpitation, trembling										
1. Project managers	**228**	**2.927**	**0.849**	**13.497**	**0.000**	**0.000**	**Arch**	**−1.348**	**0.165**	**0.000**
							CW	−0.273	0.111	**0.014**
2. Architects	**76**	**4.275**	**1.641**				**Eng.**	**1.042**	**0.229**	**0.000**
							QS	1.042	0.241	**0.000**
							CW	1.075	0.161	**0.000**
							Others	1.042	0.268	**0.000**
3. Engineers	48	3.233	0.000							
4. Quantity surveyors	41	3.233	0.000							
5. Construction workers	276	3.199	1.601							
6. Others	30	3.233	0.000							
Total	702	3.233	1.663							

Note: QS - Quantity surveyors; Arch - Architects; CW - Construction workers; Eng - Engineers.
Figures **bolded** – Significant between-group differences

Table 3.6 One-way between-groups ANOVA for stress of construction personnel in different age groups.

Age groups	N	Mean	SD	F (ANOVA)	Sig (ANOVA)	Sig (Levene)	Group	Mean diff.	SD	Sig
Work stress										
1. 29 or below	163	0.620	4.401	**2.754**	**0.042**	**0.812**				
2. 30–39	183	0.450	5.097							
3. 40–49	242	0.517	4.161							
4. 50 or above	114	−0.787	4.147				**≤29**	**−1.407**	**−2.482**	**0.010**
							29–39	**−1.237**	**−2.290**	**0.021**
							40–49	**−1.304**	**−2.306**	**0.011**
Total	702	0.313	4.492							
Physical stress										
1. 29 or below	163	25.037	4.307	1.958	0.119	0.027				
2. 30–39	183	24.084	4.512							
3. 40–49	242	23.938	5.224							
4. 50 or above	114	24.249	4.198							
Total	702	24.283	4.688							
Emotional stress										
1. 29 or below	163	31.631	3.210	2.350	0.071	0.787				
2. 30–39	183	31.632	3.237							
3. 40–49	242	31.014	3.407							
4. 50 or above	114	30.902	3.130							
Total	702	31.901	3.284							

Note: **Bolded** figures – Significant between-group differences

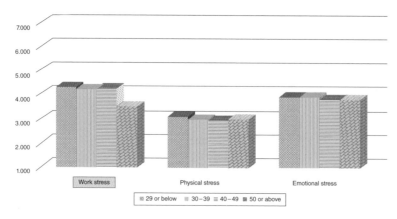

Figure 3.9 Stress of construction personnel in different age groups. Note: The −6 to 6 scale of work stress is converted to 1−7 for the sake of comparison with physical stress and emotional stress. Highlighted item—Significant differences revealed in the one-way ANOVA (refer to Table 3.6).

Table 3.7 One-way between-groups ANOVA for stress of construction personnel by gender.

Gender groups	N	Mean	SD	F (ANOVA)	Sig (ANOVA)	Sig (Levene)
Work stress						
1. Male	625	0.413	4.375	3.476	0.063	0.254
2. Female	71	−0.635	5.400			
Total	696	0.306	4.497			
Physical stress						
1. Male	625	24.224	4.667s	1.292	0.256	0.700
2. Female	71	24.892	4.905			
Total	696	24.292	4.693			
Emotional stress						
1. Male	625	31.257	3.226	**4.426**	**0.036**	**0.419**
2. Female	71	32.110	3.363			
Total	696	31.344	3.248			

Note: **Bolded** figures - Significant between -group difference.

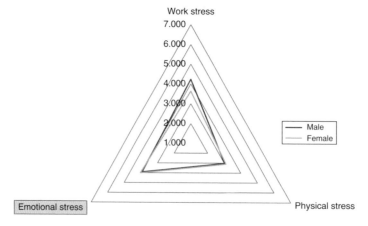

Figure 3.10 Stress of construction personnel by gender. Note: The −6 to 6 scale of work stress is converted to 1−7 for the sake of comparison with physical stress and emotional stress. Highlighted item —Significant differences revealed in the one-way ANOVA (refer to Table 3.7).

Table 3.8   Correlation coefficients of the three kinds of stress.

Kinds of stress	Work stress	Physical stress	Emotional stress
Work stress	1		
Physical stress	0.117*	1	
Emotional stress	0.246*	0.438*	1

Note: * Correlation is significant at the 0.01 level (2-tailed).

3.3.1.4 Stress by gender

A further one-way between-groups ANOVA was conducted to explore the different levels of work, physical and emotional stress experienced by male and female personnel (see Table 3.7). The significance values of work, physical and emotional stress in Levene's test (0.254, 0.700 and 0.419, respectively) were all greater than 0.05, indicating no violation of the assumption of homogeneity of variance. There was a significant difference at the $p < 0.05$ level in the emotional stress scores for the two genders ($F = 4.426$, sig < 0.05).

This result indicates that the emotional stress suffered by the male staff was significantly lower than that of their female counterparts (see Figure 3.10). In fact, women have been shown to be more susceptible to emotional stress symptoms such as anxiety and depression, which may be due to their different biological mechanisms, such as hormonal secretions and visual arousal (Cahill 2006; Felmingham et al. 2012).

3.3.1.5 Pearson correlation between stress types

Finally, in order to understand the relationships between the three types of stress for this sample of construction personnel, a correlation analysis was conducted (see Table 3.8). Significant positive correlations emerged between work and physical stress ($r = 0.117$, $p = 0.01$), work and emotional stress ($r = 0.246$, $p = 0.01$) and physical and emotional stress ($r = 0.438$, $p = 0.01$). These close interrelationships have been reviewed in a number of previous studies (see Section 3.2).

3.4   Case Studies

3.4.1   Public Toilet Construction

This project was located in a town centre near a public vegetable market and aimed at renovating an existing public toilet. The goal was to enlarge capacity and provide more advanced facilities. The scope of works covered the demolition of the existing structure and the construction of a new one-storey public toilet building with additional cubicles and renewed sanitary fittings such as water closets, hand-wash basins and urinals. It was classified by the firm as a minor and simple project. After the tendering process, the contract was awarded to a large construction firm who would be paid the contract sum of US$5.5 million over the contract period of 360 days. In order to ensure the convenience of public access to the nearby vegetable market, construction work was only allowed to be undertaken within a designated period of each day.

Construction personnel involved in this project suffered from different level and type of stress. To collect comprehensive qualitative data on their stress levels, the construction personnel, including the project manager, architects, engineers, quantity surveyors and construction workers, were interviewed individually. They were asked questions about their work, physical and emotional stress. The descriptive data was then analysed to identify, collect and summarise key words and phrases (see Table 3.9).

At the beginning of the process, the project manager thought that the construction was simple and did not require any specialist trades such as pile construction. No particular training was required. Skilled workers, such as concreters and steel benders, found the structure and design offered little challenge and usually carried out the work without paying much attention to the relevant drawings. Both the project manager and workers believed that they could complete the work ahead of schedule and hence spent more time on other issues.

Even though the construction was not complicated, a proper inspection procedure should be followed in all public works; such as a site inspection by a consultant or the client's representative. However, the engineering consultant seldom visited the site for this purpose. He assumed that it would be the same for him to conduct the inspection based on contractor's documents. Although the engineer noticed that there was insufficient fencing around the working area and reported this to the project manager, the manager did not take any immediate action and simply replied that his workload was heavy and he would attend to it when he is available. Meanwhile, consideration of the long period of time spent on purchasing new equipment, due to the long distance between the site and the nearest town, the project manager allowed the workers to carry out their work with inadequate safety equipment. He planned to purchase newer equipment later on.

In view of the small scale and simplicity of the project, the project manager did not make much effort to bring together the different parties and did not issue regular reports to the consultants. Therefore, the architect was forced to take the leading role in information processing and construction management.

As this public toilet renovation project was considered a minor piece of work, the construction personnel involved, including the project manager and construction workers, did not experience too much *work stress* in terms of the number of tasks, workload and the complexity of their work (see Table 3.9). The project manager stated that he could 'totally handle the simple tasks', and the engineer agreed that his 'workload was not heavy'. Moreover, the construction worker also reported that he 'often completed tasks early and took a rest'. In line with the results of the survey reported earlier, older staff are less likely to suffer from work stress. For example, the quantity surveyor in this case, who had over 20 years of experience, described his work on this project as 'simple' and explained that he could 'totally handle such an easy job'. The inadequate challenges

Table 3.9   Stress experiences of construction personnel involved in the public toilet project.

Stress	Construction personnel
Work stress	PM: The public toilet renovation project is a minor piece of work. I can totally **handle** it. Honestly, I don't need to spend too much time on the **simple tasks**. Eng: My **workload is not heavy**. I do not have to visit the project site regularly, because I believe that the workers can complete such a simple project according to the drawings and instructions. QS: Since I have *over 20 years' experience*, the calculation of this project is simple. I can totally **handle** such an easy job. CW: The construction of the public toilet is **not complicated**. *As an experienced worker*, I often complete my tasks early and can then take a rest. I don't even bring along the drawings as the work is **simple**.
Emotional stress	PM: This minor work, the construction of a public toilet, is a simple project. I can handle it easily. Hence, I really **enjoy** my job and feel **relaxed** during this project. Arch: Site management is so poor for this project. The project manager does not take the project seriously and failed to submit progress reports on time. I feel **anxious** about the final quality of the public toilet. Eng: I did not inspect the project site regularly. However, when I found out about the mismatch between the drawings and construction works, I **lost my temper** and spoke foul language directly in front of the construction workers. QS: I am **bored** at my work. As this is a minor project, the calculations involved in the public toilet are simple but repetitive. There are only a few major facilities that need to be calculated, such as toilet cubicles and sanitary fittings. It is a bit frustrating that I am assigned such a simple task. CW: The construction of the public toilet is a simple job for me. But I am **worried** about my next job as this is a rather short project. I need to find another job to support my family.
Physical stress	PM: My health is really good. I have had **no** significant **physical problems (such as headaches or musculoskeletal pain)** during the project. Eng: When I noticed that there was insufficient fencing fixed around the work area, I felt a **sudden stomachache.** CW: The construction work is simple . . . However, the project period is too short. I experienced sleep disorders and thought at night about finding a new job.

Note: PM = Project manager; Arch = Architect; Eng = Engineer; QS = Quantity surveyor; CW = Construction worker.

involved in such easy job tasks can result in understimulation, which leads to boredom.

As the project was simple, the project manager did not experience negative emotional symptoms. In fact, he said that 'I really enjoyed job and felt relaxed during this simple project'. However, the task simplicity made the quantity surveyor feel bored and frustrated with the calculation and estimation of the toilet cubicles and sanitary facilities. The engineer and the architect also reported emotional stress. At the beginning of the project, the engineer was relaxed and did not inspect the project site regularly. However, he lost his 'temper when I found out that there was the mismatch between the drawings and construction works'. The architect was concerned about the site management and 'felt anxious about the final quality of the public toilet', when she found out that the project manager was not taking the project seriously. The construction worker also mentioned that he was worried about his next job because this was a rather short project. The worker would need to 'find another job to support his family'.

In the process of the renovation project, the project manager did not experience *physical stress* or related problems such as headaches and musculoskeletal pain. However, the engineer felt a sudden stomachache when he noticed the problem with the perimeter fencing. The construction worker also 'experienced sleep disorders and thought at night about finding a new job'.

3.4.2   Highway Project

The highway project mainly consisted of constructing a 4.5 km dualled three-lane trunk road with a 3.7 km tunnel linking up four main districts in a developed, bustle and hustle city. In addition, it also embraced a diverse range of associated works, including the modification of an existing road nearby, construction of administration and ventilation buildings, the installation of noise barriers and semi-enclosure noise barriers along the main highway, drainage works and landscaping works. The aim was to separate traffic from the commercial centres and alleviate the heavy congestion between the four districts. It was expected that the completion of the project would generate seven practical benefits to the public, including a faster route, reduced noise, better air quality, enhanced water quality, more greenery in the area, a more sustainable city and more job opportunities (Highways Department 2014).

The Highways Department and the consultant were responsible for the project management and consultancy services, respectively. The estimated project cost was US$3.6 billion over an eight-year period. To facilitate the project, a temporary routing arrangement was implemented in the affected districts, which further exacerbated local traffic congestion.

The project manager took responsibility for leading the team to achieve the project goals (i.e., completing on time, controlling costs, ensuring quality and enhancing safety and environmental performance). He was engaged in various tasks including communication and coordination with project team members, exploration of innovative construction technologies and monitoring costs. The highway project adopted a comprehensive management system, emphasised good-quality site management and project culture and mandated a safe environment. However, the project manager still complained that it was very difficult to coordinate the different parties, as this project involved over 6000 construction personnel. Lots of subcontractors were engaged, who seldom fully complied with the management system. To ensure high-quality site management, project managers had to communicate with the project team very frequently and pay close attention to practical implementation. In addition, due to the huge amount of investment in the project, it was very challenging for the project manager to control costs and commit purchase orders appropriately without additional claims and wastage. As the four main districts affected by the highway project were the most prosperous and busy areas in the city, inappropriate arrangements for transporting material might also have seriously exacerbated the congestion. Therefore, the project manager was also required to minimise any disturbance to these districts during the construction process.

The project was also challenging for the architects. They not only aimed to design a functional expressway, but also to upgrade working and living standards of the locals through means such as installing noise barriers, conducting landscaping works to enhance the environment and designing ventilation-conducive buildings to improve air quality. To meet all these requirements, the architects had to prepare a design that would optimally balance all these elements. In practice, they often needed to seek comments from various other parties including the client, different consultants and experts such as air quality specialists and transportation engineers, end users such as residents and motorists, in order to arrive at a final design decision which balanced the interests of all parties.

There were lots of buildings in the four districts which would affect and be affected by the highway project. For instance, the foundations of existing buildings will be impaired if the engineers on the project had failed to properly deal with soil conditions. In fact, the structural engineers complained that the architects had produced incredibly complicated schemes that were almost practically and technically unfeasible. As a result, they had to make huge efforts, often using trial and error methods, to produce reliable structural designs. The amount of work involved in this caused them significant frustration.

Quantity surveyors took responsibility for all cost-related issues in the project, including cost planning, contractual arrangements, assessment of interim payments, valuation of variations and evaluation of claims. Since this was a giant project involving different types of construction (such as the expressway, noise barriers and buildings), the quantity surveyors reported that they had to manage a huge amount of cost and contractual issues such as estimation and valuation of thousands of variations. Most of these tasks were totally new to them, so they were fully occupied by their work and most of the time felt that it exceeded their abilities in terms of both quantity and quality.

For safety reasons, all contractors were required to provide construction workers with a full set of personal protective equipment (PPE), to employ sufficiently skilled workers, to provide appropriate specialist training, to organise daily morning exercise sessions and so on. Although the provision of safety equipment can protect construction workers from injuries or even death, learning how to properly use advanced PPE may be difficult for them. In addition, the workers reported that the PPE provided did not fit them very well and they experienced problems, such as fogging and scratching of the eyewear, which reduced their productivity and extended their working hours. They thought that the training had made a very limited contribution to resolving these problems. In fact, participating in a plethora of training activities and the daily morning exercise session increased their workload.

Table 3.10 Stress experiences of construction personnel involved in the highway project.

Stress	Scripts of construction personnel
Work stress	PM :You cannot imagine **how much** work I have to handle and some of the tasks, including land reclamation along the harbour front, relocation of transportation hub and systems (e.g. ferries, mass transit railway, buses, minibuses, taxis) in the town centre, liaison with multiple stakeholders and so on, are really **difficult**. For example, it is almost impossible to precisely assess variation claims for such a complicated project. Arch: I have made lots of attempts at producing an optimal design scheme for this **complex project**. The expected design should meet different requirements, such as sound-absorbing panels for reducing noise, landscaping works for enhancing greenery and ventilation systems for improving air quality. To understand the requirements of different parties, I often meet with the client, different consultants and local residents. However, it is **not easy** for us, particularly the younger architects, to handle such a comprehensive project and satisfy all parties. Eng: The project is located near the harbour front and business districts. I have to consider the soil bearing capability, the underground water and the foundations of nearby high-rise buildings . . . There are **too many things** that I have to consider and none of them are **simple**. QS: I have to **set cost plans, prepare bills of quantities and evaluate various variations** for different parts of the construction, as this project is complicated and has changed many times according to different requirements. The **continuous variation claims** result in work overload. CW: A long duration is inevitable for this complex project. However, the additional safety equipment often further slows down our work **speed** and reduces my **productivity**, which means I have to **work for a long time every day**.
Emotional stress	PM: Frankly, I often **lose my temper** when my subordinates make stupid mistakes. Sometimes, I will feel **emotionally drained** after coordinating my team members and subcontractors. Arch: It is impossible to generate a design that can optimally meet all requirements. It particularly **frustrates** me and causes me great **unhappiness** when my design is repeatedly rejected by stakeholders, whether they are technical specialists or local residents. Eng: I feel **fatigued** and **anxious** about my design scheme, as it is not perfectly reliable. All I can do is to recalculate again and again, which just makes me feel **used up**. QS: It is **hard to keep calm** when you have lots of difficult jobs to do within a limited time. I feel very **tired** every weekday and even at weekends. CW: I get **tired** because of the heavy physical work. Sometimes, I have also **worried** about my safety.

Table 3.10   *(Continued)*

Stress	Scripts of construction personnel
Physical stress	PM: The project duration is tight for such complicated work. I **do not want to eat** at all at each milestone of the project. I feel a **serious headache** every time I hear about a new problem.
	Arch: I get **serious migraines**, as I can't stop thinking about my design scheme after each meeting with different parties. In addition, since I have to sit for a long period at work, my **muscles are really painful**.
	Eng: Due to my limited experience, I am not very sure whether my design scheme is reliable, which heavily **impairs my sleep quality** and causes me nightmares accompanied by **sweating and palpitations**.
	QS: I **lose my appetite**, especially when new variations claims are initiated. Sometimes, I cannot **sleep** well after a busy working day.
	CW: Due to the repetitive and physically demanding tasks involved, I often suffer from **skeletal pain** during and after work. I hope that I can rest after a long working day. However, I also suffer from sleep **disorders**, especially during the project completion stage when the pace is really fast. This is a vicious cycle.

Note: PM = Project manager; Arch = Architect; Eng = Engineer; QS = Quantity surveyor; CW = Construction worker.

Understandably, the construction personnel involved in this project experienced stress, as all of them faced various difficulties that could serve as stressors. To collect comprehensive qualitative data on their stress levels, the construction personnel, including the project manager, architects, engineers, quantity surveyors and construction workers, were interviewed individually. During the case study, they were asked questions about their work, physical and emotional stress. The descriptive data was then analysed to identify, collect and summarise key words and phrases (see Table 3.10).

This large-scale highway project involved a lot of tasks, resulting in heavy *work stress*. The project manager was kept busy handling a mass of tasks, including 'land reclamation along the harbour front, relocation of transportation hub and systems (e.g. ferries, mass transit railway, buses, minibuses, taxis) in the town centre, liaison with multiple stakeholders and so on'. The architects, especially the younger one, also described making a lot of effort to meet the different requirements and finding it difficult to satisfy all parties. Due to the complicated soil conditions and the existence of nearby buildings, the engineers had to 'consider the soil bearing capability, the underground water and the foundations of nearby high-rise buildings' which resulted in too high a workload. The quantity surveyor complained that he needed to 'set cost plans, prepare bills of quantities and evaluate various variations', which also led to a feeling of overload. On the other hand, the construction worker reported that 'the additional safety equipment often further slows down his work speed and reduces his productivity', which extended the time spent at work. These tasks were considered both qualitatively and quantitatively demanding by these construction personnel, inducing stress.

The emotional stress of these personnel mainly manifested itself in reactions such as losing one's temper, feeling drained, frustration, unhappiness, feeling used up and being fatigued. Again, these are very similar reactions to those reported in the questionnaire study. The project manager admitted that he had 'often' lost his temper and that he felt 'emotionally drained after coordination meetings with team members and subcontractors'. As it was impossible to meet all the requirements simultaneously, the architect was particularly 'frustrated and unhappy' when his design was repeatedly rejected by different stakeholders. The engineer also indicated that he felt 'fatigued and anxious' and compelled to 'recalculate again and again'. The complicated design and numerous variations also caused fatigue for the quantity surveyors, who commented that 'it is hard to keep calm when you have lots of difficult jobs to do within a limited time'. Due to the long working days, the construction workers became tired and were also worried about their safety in the risky environment.

All personnel reported that they suffered from *physical stress* such as headache, sleep disorders, loss of temper, loss of appetite, musculoskeletal pain, sweating and palpitations. The project manager said that he 'did not want to eat at all at each milestone of the project' and got headaches when encountering new problems. Due to the different expectations of stakeholders, the architect described suffering from 'serious migraines' caused by the fact that he could not 'stop thinking about my design scheme after each meeting with different parties'. In addition, he also mentioned that sitting for the whole working day caused him muscular pain. Due to their limited experience, the younger engineers were stressed to the point that the quality of their sleep was impaired, as well as suffering from sweating and palpitations. In an echo of the findings of the questionnaire survey reported earlier, the quantity surveyor reported loss of appetite 'especially when new variations or claims are initiated', and also felt unable to sleep after a busy day. Due to the 'repetitive and physically demanding tasks', some construction workers suffered from musculoskeletal pains and sleep disorders.

3.5   Practical Implications

Based on this extensive literature, three types of stress have been identified, including work stress arising from P–E misfit, physical stress from the homeostatic effect and emotional stress governed by the automatic limbic-generated process. Actual ability plus the expectations and valence of job performance or outcomes can affect stress of construction personnel. A mismatch between ability and work demands can induce individual physical and/or emotional problems, with physical and emotional stress levels being closely associated in daily work and life. The three types of stress are, in fact, part of a closely interrelated mechanism.

The RO–BO scale (i.e., the deviation between expected and actual abilities) together with lists of physical (such as pain, appetite loss and high blood pressure) and emotional (such as worry, fatigue and unhappiness) stress symptoms were used to develop a survey measuring the work, physical and emotional stress levels of a sample of construction personnel. The results show that their work stress was within the optimisation zone and all physical stress items were scored below neutral. However, the number of projects that they were required to handle was the most difficult issue for them, with musculoskeletal pain and appetite loss the most serious physical stress symptoms experienced in their daily lives. Although these construction personnel had sufficient ability and were physically healthy enough to do their work, they still felt used up, anxious, fatigued and worried. The results of a correlation analysis further identified the close interrelationships between these three types of stress. Therefore, we should not simply conclude that construction personnel are suffering no distress in terms of their work and physical health. On the contrary, they still need to take care of their work environment and physical well-being, especially the number of projects to be handled, musculoskeletal wellness, problems of sweating and palpitation and appetite in order to avoid a negative impact on their emotional stress levels.

On the other hand, the study demonstrates that there was no significant difference in work and emotional stress levels between construction personnel in different disciplines such as project managers, architects and engineers. However, the physical stress levels of quantity surveyors and construction workers were found to be significantly higher than those of project managers. Due to the physically demanding role of construction workers, considerable attention has been paid in recent years to providing them with health and safety training so as to enhance their physical well-being (Wilkins 2011). However, our results shed light on the importance of promoting physical health care to quantity surveyors as well, although they are not required to work on site as frequently as other personnel.

In terms of age, there was no significant difference in physical and emotional stress levels for different age groups, but the work stress of personnel aged 50 or above was significantly lower than that of younger staff. In addition, construction personnel aged 50 or above reported a negative mean score for work stress. This means that their actual ability was higher than the demands of their work, especially when compared with the group aged 29 or below. It is also interesting that female construction personnel reported more emotional stress than males.

To further explore the stress experiences of construction personnel, a case study was also carried out. A systematic contextual analysis of interview data identified 10 symptoms across the three stress categories, namely work (qualitative and quantitative work demands), physical (sleep disorders, stomach pain, increased heart rate, tiredness and headaches) and emotional (worry, anxiety and losing one's temper). These reported experiences mirror the issues identified in our review of the quantitative evidence.

3.6   Summary

This chapter has reviewed the concepts and evidence surrounding the stress levels of construction personnel in terms of work, emotional and physical stress. The three types of stress are explained by different. A research project including both questionnaire survey and case study was conducted to investigate the stress of construction personnel. Three types of stress, namely work, emotional and physical stress, were identified, consistent with the theoretical framework. The results indicated that the number of projects worked on, musculoskeletal pain, appetite loss, feeling used up, anxiety and worry were the areas of most importance to these construction personnel. A comparison between different personnel groups in terms of discipline, age and gender was also conducted. Quantity surveyors and construction workers faced more physical stress than project managers, as did younger personnel. Female staff were found to suffer more emotional stress than male staff. The data also supported a significant interrelationship between work, physical and emotional stress. To validate the quantitative data, a case study was also conducted to understand the real-life impact of the stress suffered by construction personnel.

References

Aronson, A.E. and Kafry, D. (1998) *Managing Burnout*. London, Kogan Page.

Atkinson, J.W. (1957) Motivational determinants of risk-taking. *Psychological Review*, 64, 359–372.

Barrett, L.F., Mesquita, B., Ochsner, K.N. and Gross, J.J. (2007) The experience of emotion. *Annual Review of Psychology*, 58, 373–403.

Bowles, D. and Cooper, C.L. (2009) *Employee Morale: Driving Performance in Challenging Times*. Basingstoke: Palgrave Macmillan.

Cahill, L. (2006) Why sex matters for neuroscience. *Nature Reviews Neuroscience*, 7, 477–484.

Carlson, N.R. (2010) *Psychology: The Science of Behavior*. New Jersey: Pearson.

Carroll, J. and White, W. (1982) *Theory Building: Integrating Individual and Environmental Factors with an Ecological Framework*. Beverly Hills, CA: Sage.

Chan I.Y.S., Leung M.Y, Yu, S.W. (2012) Managing Stress of Hong Kong expatriate construction professionals in Mainland China: A focus group studies to exploring individual coping strategies and organizational support, *Journal of Construction Engineering and Management*, ASCE, 138(10), 1150–1160.

Chrousos, G.P. (2009) Stress and disorders of the stress system. *Nature Reviews Endocrinology*, 5, 374–381.

Chrousos, G.P. and Gold, P.W. (1992) The concepts of stress and stress system disorders: Overview of physical and behavioral homeostasis. *Journal of American Medical Association*, 267(9), 1244–1252.

Chrousos, G.P. and Kino, T. (2007) Glucocorticoid action networks and complex psychiatric and/or somatic disorders. *Stress*, 10(2), 213–219.

CIOB (2006) Occupational stress in the construction industry. *CIOB Published National Stress Survey Results*. Retrieved from http://www.ciob.org.uk/resources/publications

Cohen, S., Tyrrell, D.A.J. and Smith, A.P. (1991) Psychological stress and susceptibility to the common cold. *New England Journal of Medicine*, 325, 606–612.

Cordes, C.L. and Dougherty, T.W. (1993) A review and an integration of research on job burnout. *Academy of Management Review*, 18(4), 621–656.

Cousins, R., MacKay, C.J., Clarke, S.D., Kelly, C., Kelly, P.J. and McCaig, R.H. (2004) 'Management Standards' work-related stress in the UK: Practical development. *Work and Stress*, 18(2), 113–136.

Cox, T. (1993) *Stress Research and Stress Management: Putting Theory to Work*. HSE Contract Research Report, No. 61/1993.

Daniels K. and Guppy A. (1994) Occupational stress. *Human Relations*, 47(12), 1523–1544.

Descartes, R. (1967) The passions of the soul. In E.S. Haldane and G.R.T. Ross (eds), *Philosophical Works of Descartes*. New York: Dover.

Diefendorff, J.M. and Croyle, M.H. (2008) Antecedents of emotional display rule commitment. *Human Performance*, 21, 310–332.

Edwards, J.R. (1996) An examination of competing versions of the person–environment fit approach to stress. *Academy of Management Journal*, 39, 292–339.

Erez, A. and Isen, A.M. (2002) The influence of positive affect on components of expectancy motivation. *Journal of Applied Psychology*, 87, 1055–1067.

Fayek, A.R., Revay, S.O., Rowan, D. and Mousseau, D. (2006) Assessing performance trends on industrial construction mega projects. *Cost Engineering*, 48(10), 16–21.

Felmingham, K.L., Tran, T.P., Fong, W.C. and Bryant, R.A. (2012) Sex differences in emotional memory consolidation: The effect of stress-induced salivary alpha-amylase and cortisol. *Biological Psychology*, 89(3), 539–544.

French, J.R.P. and Caplan, R.D. (1972) *Organizational Stress and Individual Strain: The Failure of Stress*. New York: AMACOM.

French, J.R.P., Rogers, W. and Cobb, S. (1974) *Adjustment as a Person–Environment Fit, Coping and Adaption: Interdisciplinary Perspectives*. New York: Basic Books.

French, J.R.P., Caplan, R.D. and Harrison, R.V. (1982) *The Mechanisms of Job Stress and Strain*. New York: Wiley.

Freudenberger, H.J. (1983) Burnout: Contemporary issues, trends, and concerns. In B.A. Farber (ed.), *Stress and Burnout in the Human Services Professions*. New York: Pergammon Press, 23–28.

Gmelch, W.H. (1982) *Beyond Stress to Effective Management*. New York: Wiley.

Gilbey, M.P. and Spyer, K.M. (1993) Essential organization of the sympathetic nervous system. *Baillieres Clinical Endocrinology Metabolism*, 7(2), 259–278.

Hair, J.F.J., Anderson, R.E., Tatham, R.L. and Black, W.C. (1998) *Multivariate Data Analysis*. New Jersey: Prentice Hall.

Harrison, R.V. (1978) *Person–Environment Fit and Job Stress*. New York: Wiley.

Haynes, N.S. and Love, P.E.D. (2004) Psychological adjustment and coping among construction project managers. *Construction Management and Economics*, 22, 129–140.

Health and Safety Executive (2012) *Stress and Psychological Disorders 2011/2012*. Retrieved from http://www.hse.gov.uk/statistics/causdis/stress/stress.pdf

Holsboer, F. (1999) The rationale for corticotropin-releasing hormone receptor (CRH-R) antagonists to treat depression and anxiety. *Journal of Psychiatric Research*, 33(3), 181–214.

Hung, C.I., Liu, C.Y. and Wang, S.J. (2008) Precipitating or aggravating factors for headache in patients with major depressive disorder. *Journal of Psychosomatic Research*, 64, 231–235.

Janssen, P.P.M., Schaufeli, W.B. and Houkes, I. (1999) Work-related and individual determinants of the three burnout dimensions. *Work and Stress*, 13(1), 74–86.

Kominis, G. and Emmanuel, C.R. (2006) The expectancy-valence theory revisted: Developing an extended model of managerial motivation., *Management Accounting Research*, 18(1), 49–75.

Kristof-Brown, A.L., Zimmerman, R.D. and Johnson, E.C. (2005) Consequences of individuals' fit at work: A meta-analysis of person–job, person–organizational, person–group and person–supervisor fit. *Personnel Psychology*, 58, 281–342.

Kventnansky, R., Sabban, E.L. and Palkovits, M. (2009) Catecholaminergic systems in stress: Structural and molecular genetic approaches. *Physiological Reviews*, 89, 535–606.

Kyrou, I. and Tsigos, C. (2009) Stress hormones: Physiological stress and regulation of metabolism. *Current Opinion in Pharmacology*, 9(6), 787–793.

Lazarus, R.S. and Folkman, S. (1984) *Stress Appraisal and Coping*. New York: Springer.

LeDoux, J.E. (2000) Emotion circuits in the brain. *Annual Review of Neuroscience*, 23, 155–184.

Lee, S. (2007) Vroom's expectancy theory and the public library customer motivation model. *Library Review*, 56(9), 788–796.

Leung, M.Y. (2004) An international study on the stress of estimators. *The Hong Kong Surveyor*, 15(1), 49–52.

Leung, M.Y., Skitmore, M., Chan, I.Y.S. (2007) Subjective and Objective Stress in Construction Cost Estimation. *Journal of Construction Management and Economics*, 25(10), 1063–1075.

Leung, M.Y., Chan, Y.S. and Olomolaiye, P. (2008a) Impact of stress on the performance of construction project managers. *Journal of Construction Engineering and Management*, 134(8), 644–652.

Leung, M.Y., Chan, Y.S., Chong, A., Sham, J.F.C. (2008b) Developing structural integrated stressors-stress models for clients' and contractors, cost engineers. *Journal of Construction Engineering and Management*, ASCE, 134(8), 635–643.

Leung, M.Y., Chan, Y.S., Yu, J.Y. (2008c) An Integrated Model for the Stressors and Stresses of Construction Project Managers, *Journal of Construction Engineering and Management, ASCE*, 135(2), 126–134.

Leung, M.Y., Chan, I.Y.S., Yuen, K.W. (2010) Impacts of stressors and stress on the injury incidents of construction workers in Hong Kong. *Journal of Construction Engineering and Management, ASCE*, 136(10), 1093–1103.

Leung, M.Y., Chan, Y.S.I. and Chen, D. (2011) Structural linear relationships between work stress, burnout, physiological stress and performance of construction project managers. *Engineering, Construction and Architectural Management*, 18(3), 312–328.

Leung, M.Y., Chan, I.Y.S., Yu, J.Y. (2012) Preventing construction worker injury incidents through the management of personal stress and organizational stressors, *Journal of Accident Analysis and Prevention*, 48, 156–166.

Lewin, K. (1936) *Principles of Topological Psychology*. New York: McGraw-Hill.

Li Calzi, S., Farinelli, M., Ercolani, M., Alianti, M., Manigrasso, V. and Taroni, A.M. (2006) Physical rehabilitation and burnout: Different aspects of the syndrome and comparison between healthcare professionals involved. *Europa Medicophysica*, 42(1), 27–36.

Locke, E.A. (1969) What is job satisfaction? *Organizational Behavior and Human Performance*, 4, 309–336.

MacLean, P.D. (1949) Psychosomatic disease and the 'visceral brain': Recent developments bearing on the Papez theory of emotion. *Psychosomatic Medicine*, 11, 338–353.

MacLean, P.D. (1952) Some psychiatric implications of physiological studies on frontotemporal portion of limbic system. *Electroencephalography and Clinical Neurophysiology*, 4, 407–418.

MacLean, P.D. (1972) Cerebral evolution and emotional processes: New findings on the striatal complex. *Annals of New York Academy of Sciences*, 193, 137–149.

MacLean, P.D. (1994) Human nature: Duality or triality? *Politics and the Life Sciences*, 13(1), 107–112.

Maslach, C., Jackson, S. and Leiter, M.P. (1996) *Maslach Burnout Inventory*. Palo Alto, CA: Consulting Psychologists Press.

McEwen, B.S. (1998) Stress, adaptation and disease, allostasis and allostatic load. *Annals of the New York Academy of Sciences*, 840, 33–44.

McEwen, B.S. (2007) Physiology and neurobiology of stress and adaptation: Central role of the brain. *Physiological Reviews*, 87, 873–904.

Mommersteeg, P.M.C., Keijsers, G.P.J., Heijnen, C.J., Verbraak, M.J.P.M. and van Doormen, L.J.P. (2006) Cortisol deviations in people with burnout before and after psychotherapy: A pilot study. *Health Psychology*, 25(2), 243–248.

Monat, A. and Lazarus, R.S. (1991) *Stress and Coping: An Anthology*. New York: Columbia University Press.

Montoro, J., Mullol, J., Jauregui, I., Davila, I., Ferrer, M., Bartra, J., der Cuvillo, A., Sastre, J. and Valero, A. (2009) Stress and allergy. *Journal of Investigational Allergology and Clinical Immunology*, 19(1), 40–47.

Murray, H.A. (1938) *Explorations in Personality*. New York: Oxford University Press.

Nadler, D.A. and Lawler, E.E. (1983) Motivation: A diagnostic approach. In J.R. Hackman, E.E. Lawler and L.W. Porter (eds), *Perspectives on Behavior in Organizations*. New York: McGraw-Hill.

Nyssen, A.S, Hansez, I., Baele, P., Lamy, M. and Keyser, V.D. (2003) Occupational stress and burnout in anaesthesia. *British Journal of Anaesthesia*, 90(3), 333–337.

Oxford Dictionary (2012) *Definition of emotion*. Retrieved from http://oxforddictionaries.com/definition/american_english/emotion?region=us

Pervin, L. (1968) Performance and satisfaction as a function of individual-environment fit. *Psychological Bulletin*, 69(1), 5–68.

Pinder, C.C. (1991) Valence-instrumentality-expectancy theory. In R.M. Steers and L.W. Porter (eds), *Motivation and Work Behavior*. New York: McGraw-Hill.

Pines, A. (1982) Changing organizations: Is work environment without burnout an impossible goal? In W.S. Paine (ed.), *Job Stress and Burnout: Research, Theory and Intervention Perspectives*. Beverly Hills, CA: Sage.

Pruessner, J.C., Dirk, H.H. and Clemens, K. (1999) Burnout, perceived stress and cortisol responses to awakening. *Psychosomatic Medicine*, 61(2), 197–204.

Reul, J.M.H.M. and Holsboer, F. (2002) Corticotropin-releasing factor receptors 1 and 2 in anxiety and depression. *Current Opinion in Pharmacology*, 2(1), 23–33.

Russell, J.A. (1991) Natural language concepts of emotion. In D.J. Ozer, J.M. Healy and A.J. Stewart, (eds), *Perspectives in Personality: Self and Emotion*. London: Jessica Kingsley.

Russell, J.A. (2003) Core affect and the psychological construction of emotion. *Psychological Review*, 110, 145–172.

Schaufeli, W. (1993) *Professional Burnout: Recent Developments in Theory and Research*. New York: CRC Press.

Shirom, A. and Ezrachi, Y. (2003) On the discriminant validity of burnout, depression and anxiety: A re-examination of the burnout measure. *Anxiety Stress and Coping*, 16(1), 83–97.

Teasdale, E.L. and Mckeown, S. (1994) Managing stress at work: The ICI-Zeneca Pharmaceuticals experience. In C.L. Cooper and S. Williams (eds), *Creating Healthy Work Organisations*. Chichester: Wiley.

Tolman, E.C. (1952) *Purposive Behavior in Animals and Men*. New York: Century.

Vroom, V.H. (1964) *Motivation and Work*. New York: Wiley.

Widen, S.C. and Russell, J.A. (2010) Descriptive and prescriptive definitions of emotion. *Emotion Review*, 2(4), 377–378.

Wierzbicka, A. (2010) On emotions and on definitions: A response to Izard. *Emotion Review*, 2(4), 379–380.

Wilkins, J.R. (2011) Construction workers' perceptions of health and safety training programs. *Construction Management and Economics*, 29(10), 1017–1026.

Wong, P.S.P., Cheung, S.O. and Fan, K.L. (2009) Examining the relationship between organizational learning styles and project performance. *Journal of Construction Engineering and Management*, 135(6), 497–507.

Yip, B. and Rowlinson S. (2009) Job redesign as an intervention strategy of burnout: Organizational perspective. *Journal of Construction Engineering and Management*, 135(8), 737–745.

4

Sources of Stress Affecting Construction Personnel

4.1   Stressors Affecting Construction Personnel

Before the creation of the term 'stressors', people usually viewed stress as unpleasant threats influencing an individual, triggering one's physical or psychological responses to these stimuli. However, these are different concepts. Later on, Selye (1956) created the word 'stressor' to distinguish stress stimulus from response. In fact, stress is the combination of stressor and stress response. Without both of these, there is no stress. Therefore, stressors are an essential component of stress. They have the potential to elicit a stress reaction. In the work context, the term 'stressor' is used to designate stimuli people face on the job that have negative physical or psychological consequences for significant proportions of people exposed to them (Ganster and Rosen 2013; Lundberg and Cooper 2010).

In 1967, Holmes and Rahe collected a sample of 2500 medical records of US sailors to investigate the associations between 43 stressful life events in the past six months and various illnesses, such as headaches, ulcers and chest pain. They rated and ranked different types of life changes based on their correlations with illness. For instance, death of spouse was ranked the highest, followed by divorce, imprisonment, death of a close family member and personal injury or illness. Holmes and Rahe's stressor list gave researchers and the public a general picture of the common stressors people face in daily life. However, their stressor list was fragmented and developed for general adults in the context of general life only.

The stressors prevalent among construction personnel may not be the same as the general public, since construction personnel are often driven by demanding tasks in various temporary projects in a dynamic industry. Based on the extensive literature on stress management in the construction industry, stressors of

Stress Management in the Construction Industry, First Edition.
Mei-yung Leung, Isabelle Yee Shan Chan and Cary L. Cooper.
© 2015 John Wiley & Sons, Ltd. Published 2015 by John Wiley & Sons, Ltd.

construction personnel can generally be divided into five major categories: *personal, interpersonal, task, organisational and physical* stressors (Gmelch 1982; Leung et al. 2005; Leung, Skitmore and Chan 2007; Leung, Chan and Olomolaiye 2008; Leung, Zhang and Skitmore 2008; Leung, Chan and Yu 2011).

4.1.1 Personal Stressors

4.1.1.1 Type A personality

Personal traits have long been recognised as a significant antecedent of individuals' stress (Caplan and Jones 1975; French and Caplan 1972; Friedman and Rosenman 1974; Sogaard et al. 2007). People's personal characteristics make some people more reactive to environmental stressors, manifesting more stress than others (Gmelch 1982). Friedman and Rosenman (1974) developed a Type A and Type B personality theory, which theorised that individuals with Type A personality are more impulsive, competitive, aggressive, hasty, time driven, abiding of timetables, preoccupied by vocational deadlines, impatient, insecure of status, generally hostile and incapable of relaxing. This is the opposite of individuals with Type B personality, who are generally more easy-going and placid (Caplan and Jones 1975; French and Caplan 1972; Friedman and Rosenman 1974).

Time is a major element of construction projects. To achieve project goals, construction personnel are often forced to compete with time throughout the project period, including the initial briefing stage (contract liaisons with clients), design stage (confirmation of outline, scheme and detailed designs by architects), tendering (selection of main contractor by entire project team) and construction (on-site operation by contractors and subcontractors). A person with Type A behaviours may be good at helping the group progress towards time goals, but this trait may also induce high levels of stress. In fact, many previous studies have confirmed that Type A people report more stress and stress-related illness (e.g., coronary heart disease) than Type B people (e.g., Caplan and Jones 1975; French and Caplan 1972; Friedman and Rosenman 1974; Leung et al. 2007; Leung, Chan and Olomolaiye 2008; Sogaard et al. 2007). Type A people are also more dominant in interpersonal relationships and impatient in competitive situations than Type B people (Yarnold and Grimm 1982). In the construction context, previous studies have found that surveyors with Type A personality are more susceptible to stress (Leung, Skitmore and Chan 2007; Leung, Chan and Olomolaiye 2008). Type A and Type B people differ in their interpretation of events, which causes them to have different types of stressors. Type B people are accustomed to interpreting events as non-stressful and thus they are less likely to suffer from stress (Hagihara et al. 1997).

4.1.1.2 Pessimism

The concept of pessimism versus optimism is generally defined as negative versus positive outcome expectancies (Scheier and Carver 1985). Pessimistic people tend to anticipate negative outcomes of events around them. They tend to attribute causes of negative events to be internal (i.e., my own fault), stable over time (i.e., going to last for a long time) and global (i.e., extending the negative events to all other areas). They are also likely to appraise events as more stressful. Because they often perceive negative outcomes, they may be less likely to put effort into altering the stressful situation via problem-focused coping strategies (see Chapter 6 for details on coping). They are rather passive in the stress management process, during which time they tend to use avoidance and escapism. Thus, they have a higher likelihood of suffering from depression and low satisfaction (Chang, Maydeu-Olivares and D'Zurilla 1995).

In contrast, optimistic people tend to perceive good outcomes of events around them. They tend to adopt more adaptive coping strategies accompanied with healthy habits. Optimists actively handle stress, rather than avoid it. Optimists are reported to suffer from less depression, stress and loneliness (Scheier and Carver 1992) and have higher life satisfaction (Hayes and Weathington 2007). Hence, compared to optimists, pessimists are more susceptible to stress.

4.1.1.3 *Work–home conflict (work–life balance)*

Work-life balance has been found to be one of the essential stressors prevailing in the construction sector (Bowen et al. 2014). Construction personnel may easily encounter work–home conflict when they fail to achieve a balance between their work and private life (Emslie, Hunt and Macintyre 2004; Leung, Skitmore and Chan 2007). In general, work–home conflict can be divided into two types: work-to-home conflict (i.e., when the work domain creates demands that interfere with the home domain) and home-to-work conflict (i.e. when the home domain creates demands that interfere with the work domain; Eagle, Miles and Icenogle 1997; Frone, Russell and Cooper 1992; Greenhaus and Parasuraman 1987; Swanson, Power and Simpson 1998). However, previous studies have shown that work-to-home conflict is more common than home-to-work conflict in the context of the construction industry (Mostert, Peeters and Rost 2011).

Work-to-home conflict can happen in various ways, including work interference with marital relationships, leisure activities, the parenting role and home management (Lingard 2004; Lingard and Francis 2005). Previous studies have claimed that work–home conflict is more prevalent among females than males (Dench et al. 2002). However, the impact of work–home conflict on males is also significant enough that one cannot afford to overlook it (Emslie, Hunt and Macintyre 2004). Hence, previous studies have identified work–home conflict as a stressor influencing construction managers and estimators (Davidson and Sutherland 1992; Djebarni, 1996; Leung et al. 2005; Leung, Skitmore and Chan 2007).

4.1.2   Interpersonal Stressors

4.1.2.1   *Poor interpersonal relationships within organisation*

It is impossible for a construction employee to construct a building by him or herself, especially for high-rise buildings in modern cities. Construction personnel often work in a team with colleagues within and outside their organisation. According to Cox and Griffiths (1995), poor relations with colleagues, supervisors and subordinates at work are important risk factors for stress-related problems. In fact, proper interpersonal management is key to achieving good performance with construction personnel (Bresnen et al. 1986; Djebarni and Lansley 1995; Mustapha and Langford 1990).

Employees in large organisations spend about four out of five working hours in interactive activities, including scheduled meetings (59%), desk work sessions (22%), unscheduled meetings (10%), telephone calls (6%) and other activities (Mintzberg 1979). Therefore, interaction plays an important role in employees' work, especially for those in managerial positions. However, poor interpersonal relationships complicate the interaction process, causing stress for construction personnel. It is believed that good relationships between project team members, including relationships with superiors, subordinates and colleagues are key factors to the well-being of individuals and organisations (Cooper 2001; Sauter and Murphy 1995).

a.   *Relationships with superiors.* Superiors are often one of the most significant sources of stress for employees because employees who perceive their superiors as untrustworthy and unfriendly normally report more job pressure (Buck 1972). In addition, most people spend a great deal of time worrying about what their boss thinks of them and how their work is being evaluated (Fiedler and Garcia 1987). In fact, performance appraisal can often be subjective and significantly influenced by interpersonal relationships (Robbins and DeNisi, 1998; Varma, Denisi and Peters 1996). Hence, poor relationships with superiors can also be stressful for construction personnel.

b.   *Relationships with subordinates.* Supervising and maintaining good relationships with subordinates is one of the major responsibilities of superiors. An untrusted superior–subordinate relationship may cause difficulty in information processing, task delegation and task execution, which would all be detrimental to project performance. Previous studies have found that different leadership styles have different impacts on subordinates' stress levels. For instance, transformational leadership, which occurs when leaders focus on fostering the higher order intrinsic needs of their followers in priority to short-term needs and when followers identify the needs of the leader, can help mitigate subordinates' stress by stimulating their intellectual abilities. In contrast, transactional leadership, which occurs when leaders focus on satisfying the extrinsic needs of their subordinates and when subordinates do what the leader asks in return, tends to induce more stress among subordinates (Bass 1999; Seltzer, Numberof and Bass 1989).

c. *Relationship with colleagues.* In addition to relationships with superiors and subordinates, relationships with colleagues who often execute tasks together are also essential. It is easy to understand how poor relationships with colleagues can give rise to stress and unsatisfied needs for social association, thereby resulting in poor work cooperation. On the other hand, this kind of interpersonal relationship is one of the most important types of social support, which is usually a source of resilience to stress (Beehr 1995; Kahn and Byosiere 1992).

Interpersonal network of a project are like a complicated, invisible truss of a building that is constructed on the supports of various individuals with different personal traits. The performance of a building is highly dependent on this truss: the performance of an interpersonal network is based on individuals' different traits (supporting members) and the different ways that they interact with each other (joints). As mentioned above, people with Type A personality are abrasive and hard-driving and they are more likely to create stress for those around them. These abrasive people often ignore the interpersonal aspects of feelings and sensibilities of social interaction. All in all, studies have found that poor relationships with supervisors, subordinates and colleagues are significant antecedents of stress for various construction personnel, such as construction project managers, engineers and estimators (e.g., CIOB 2006; Davidson and Sutherland 1992; Leung, Skitmore and Chan 2007; Leung, Zhang and Skitmore 2008; Loosemore and Waters 2004; Sutherland and Davidson 1989).

4.1.2.2 *Distrust in project teams*

When playing a role in any construction project team, construction personnel interact not just with their colleagues in the same organisation, but also with many different personnel outside the organisation in both formal and informal capacities (e.g., interactions between project team members from developers, contractors and consultant firms). However, project team members may have differing values and professional disciplines or even conflicting interests, which can cause mistrust within the project team and stress for construction personnel (Quick and Quick 1984). One study found that distrust significantly impacts stress levels of quantity surveyors (Leung et al. 2005).

With lack of trust among project team members, problems may arise in information sharing, task delegation and cooperation. Distrust between project team members – either intra- or inter-organisation – can lengthen the information-processing period (Chinowsky, Diekmann and O'Brien 2010). Distrust increases not only individuals' workloads, but also the chances of project delay and monetary losses on the project, as well as stress and low job satisfaction down the line. Hence, to enhance trust among the project team, construction stakeholders are advised to organise team-building workshops or related activities (Ochieng and Price 2010).

4.1.2.3 Competitive teamwork

On the other hand, distrust can lead to competitive project teams. Instead of cooperating with each other to achieve mutual success, competitive project teams have a special atmosphere in which team members focus on comparison with each other (in terms of gainshare or painshare), emphasise personal achievement (in terms of personal career development) rather than team benefits and prefer to take the lead from others (in terms of making decisions that influence other team members; Wittchen et al. 2013). Competitive teamwork escalates the stress level of construction personnel and it has adverse effects on project success (Chiocchio et al. 2011).

4.1.3   Task Stressors

4.1.3.1 Work overload (quantitative versus qualitative)

Due to the dynamic nature of construction and tight project timeframes, work overload is common among construction personnel (Djebarni 1996; Leung et al. 2005; Yip 2008). Previous studies have found that work overload has a significant impact on stress in construction personnel (e.g., CIOB 2006; Janssen et al. 2001; Leung, Zhang and Skitmore 2008; Loosemore and Waters 2004; Sutherland and Davidson 1993). Work overload occurs when an individual is assigned with excessive job tasks (Gmelch 1982). In fact, both having 'too much' and 'too difficult' work can result in stress. French and Caplan (1972) distinguished between quantitative and qualitative work overload. Quantitative overload refers to 'having too much to do', while qualitative overload means work that is 'too difficult'.

Quantitative work overload occurs when an individual encounters difficulties in completing a certain number of assignments within a limited period of time (Katz and Kahn 1978; Maslach and Jackson 1984; Selmer and Fenner 2009). A study conducted by the CIOB in 2006 found that having 'too much work' was the main factor influencing stress levels of construction personnel. In fact, the working hours of construction personnel were found to be longer than professionals working in other industries, such as manufacturing and commerce (Pheng and Chuan 2006). Construction tasks involve not only daily paperwork, but also a great deal of time-consuming, ad-hoc engagements, such as site visits, meetings, phone calls and informal gatherings, which may lead construction personnel to suffer from quantitative work overload. In addition to stress, quantitative overload has been found to worsen people's health, trigger unhealthy behaviours like cigarette smoking and lower job satisfaction (French and Caplan 1970; Porter and Lawler 1968; Quinn et al. 1971).

Qualitative work overload occurs when people perceive a task as too difficult (Ahuja and Thatcher 2005; Katz and Kahn 1978; Maslach and Jackson 1984; Selmer and Fenner 2009). A manager is considered to suffer from qualitative work overload when he or she perceives difficulty in supervising, managing, training, developing and motivating subordinates (Ahuja and Thatcher 2005). In addition, although technology is beneficial to the work efficiency and productivity of individuals and organisations, there is some research showing that the rapid pace of technological change may not benefit professionals, particularly in terms of their workload on keeping pace with this ever-updating knowledge and skills, physical expenditure on equipping themselves with this new technology, personal change and so on (Ganster and Schaubroeck 1991). For instance, new technologies like expert systems for decision making (Chang and Ibbs 1990) and Building Information Modeling (BIM) for visualising 5D or 6D project plans and estimates (National BIM Standard, United States, 2012) have been spreading rapidly in the construction industry in recent years. Due to the novelty of these methods, it is impossible for a majority of construction personnel to have received training in these innovative techniques in their undergraduate professional education. With more and more clients requesting the use of innovative technologies in construction projects and as the associated changes become a part of the job, construction personnel experience qualitative overload, causing stress.

4.1.3.2 *Work underload (quantitative versus qualitative)*

Although excessive workload has negative impacts on individuals, it does not mean that the complete absence of work pressure of any sort creates a psychologically comfortable state (Keenan, Copper and Marshall 1980). Work underload occurs when people find themselves in a position where the demands of their jobs are insufficient to make full use of their skills and abilities (Leung et al. 2005; Leung, Zhang and Skitmore 2008). In other words, 'work underload' refers to a state in which people are understimulated at their work. Similar to work overload, work underload can be either quantitative or qualitative (Garfield 1995). Quantitative work underload happens when people have insufficient work. Qualitative work underload occurs when people's skills and abilities are not fully utilised, in which case they feel their work is boring and repetitive.

In the construction industry, work underload arises when construction personnel have insufficient tasks during surpluses or recessions (i.e., quantitative underload), which causes boredom and apathy. It can also happen when they are handling boring and repetitive tasks, such as too much non-challenging paperwork or repetitive construction processes (i.e., qualitative underload). Work underload has been found to have an even greater impact than work overload on people's stress levels (Garfield 1995). It is also common among construction personnel, particularly for quantity surveyors who are often required to deal with routine estimation tasks (Leung et al. 2007; Leung, Chan and Olomolaiye 2008).

4.1.3.3 *Role ambiguity*

The term 'role' refers to 'a set of norms that define behaviours appropriate for and expected of various positions within a group' (Francesco and Gold 2004). Role ambiguity arises when people have limited information and uncertainty about their responsibility, authority, performance expectations, performance evaluation, allocation of time, existing guidelines, organisational/departmental directions and policies (Handy 1985; Katz and Kahn 1978; Rizzo, House, and Lirtzman 1970).

Uncertainty can trigger stress (Beehr and Bhagat 1985; Landy 1992). Among the various types of role ambiguities, advancement ambiguity arises from uncertainty about one's job promotion or career development and it contributes significantly to stress (Carayon 1992). This is simply because losing one's job or simply anticipating losing one's job is stressful (Beehr 1995). In addition, role ambiguity has also been found to be strongly associated with increased levels of work stress and emotional stress, as well as lower job performances, job satisfaction and so on (Bhanugopan and Fish 2006; French and Caplan 1970; Kahn et al. 1964; Margolis, Kroes and Quinn 1974; Rizzo, House, and Lirtzman 1970; Yip 2008). To prevent role ambiguity, construction stakeholders should provide adequate information to construction personnel regarding their job responsibilities, authority, project roles and professional duties.

4.1.3.4 *Role conflict*

Although the term 'role conflict' seems simple, its definition differs from researcher to researcher. It can be considered as the incompatibility of role demands and conflicting expectations at work (Ilgen and Hollenbeck 1991), the dilemma of damned if you do, damned if you don't (Greenberg 1999) and the incompatible demands of multiple roles carried by the employees (Francesco and Gold 2004). Role conflict exists when an individual in a particular work role is torn by conflicting job demands or doing things he or she really does not want to do or does not think are part of the job specification (Mmaduakonam 1997). Kahn and his colleagues (1964) summarised role conflicts as falling into four types: intra-sender conflict; inter-sender conflict; person–role conflict; and inter-role conflict. Intra-sender conflict occurs when incompatible expectations are received from the same sender. Inter-sender conflict occurs when incompatible expectations are received from two or more role senders. Person–role conflict is when an individual disagrees with other role sender(s) about his or her own role expectations, while inter-role conflict exists when expectations of one role contradict that of another role. In general, role conflict can be thought of as multiple, conflicting expectations exerted on the role of an individual (Beehr 1995).

Previous studies have found that people who suffer from role conflict have higher work stress, higher emotional stress, higher physical stress, lower job performance and higher intention to resign (Fisher and Gitelson 1983; Ganster, Fusilier and Mayes 1986). In the construction context, role conflict is a significant source of stress for construction personnel who have to carry multiple and incompatible roles in dynamic and complicated construction projects (Leung et al. 2005; Leung, Skitmore and Chan 2007; Leung, Sham and Chan 2007; Leung at al. 2007c, Leung, Chan and Olomolaiye 2008; Sutherland and Davidson 1989).

Kahn and his colleagues (1964) identified six common possible causes of role ambiguity and conflict, which are applicable to the construction industry.

These causes include: (i) construction personnel work in a dynamic industry that is highly susceptible to the market economy and a majority of construction companies were streamlined during the recent recession; (ii) construction personnel are involved in complicated construction projects with multiple stakeholders and parties and they experience conflicting expectations from clients, project managers, consultants, contractors, subcontractors, direct supervisors and subordinates; (iii) construction personnel work for more than one supervisor, especially in large joint-venture projects, which often have several directors or contract managers; (iv) the need to supervise a multidisciplinary team members, including structural engineers, building services engineers and quantity surveyors; (v) construction personnel are required to make innovative and often cross-disciplinary decisions, especially since construction management and construction design are never routine tasks; and (vi) construction personnel need to coordinate tasks between various teams either within firms (e.g., the estimation department, materials departments and plants departments) or between firms (e.g., clients, contractors, subcontractors and design consultancies).

4.1.3.5 *Lack of job autonomy*

Job autonomy refers to the freedom and control that individuals have over their behaviours at work (Jackson 1989). When construction personnel perceive that they are being treated solely as an instrument or a means to an end by their supervisor, they may feel emotional stress, low morale and low self-esteem. It is indeed difficult for construction personnel to carry out their job effectively without adequate autonomy. Having to refer matters upwards when people perceive that they can deal with them adequately by themselves can not only be frustrating, but can also delay work progress. This may lead to further work stress and low job satisfaction.

In fact, autocratic supervisors – who often supervise their subordinates closely and give them less autonomy – are most frequently mentioned as a salient source of stress for employees (Djebarni and Lansley 1995). Numerous studies have found that a lack of job autonomy is directly associated with depression, low self-esteem and decreased job satisfaction. Furthermore, depression, low self-esteem and dissatisfaction have all been found to be positively related to stress (Beehr 1995; Nevels 1986). In the construction context, Leung and her colleagues (Leung, Sham and Chan 2007) have shown that construction estimators' lack of job autonomy is directly and negatively associated with stress.

4.1.3.6 *Lack of feedback from superiors*

As mentioned earlier, employees spend a great deal of time in worrying about how their superiors appraise their performance. Therefore, adequate feedback from superiors about subordinates' performance is an essential part of the management process, not only to ensure that employees perform well, but also to help them develop their career and, more importantly, to reduce their stress level. Although performance appraisal helps superiors adjust the workload and job nature of their subordinates, this process often induces stress in employees (Beehr and Jex 2001).

Instead of giving a poor performance grade to a subordinate at the end of the task, many potential problems can be prevented through a regular feedback process. For instance, junior project coordinators may continuously be stressed and worry about the appropriateness of the construction programmes they draft because inappropriate programmes could result in serious problems, such as a delay in completing a project and additional costs. This stress can be alleviated if superiors can provide useful comments, such as identifying potential clashes, so that the project coordinators can have confirmation of the task sequences, directions and expectations. In fact, a lack of feedback has been found to be a significant stressor of quantity surveyors in the construction industry (Leung, Sham and Chan 2007). To prevent distress among construction personnel, a competent superior should develop effective means to provide timely, regular and adequate feedback about their subordinates' tasks and performance.

4.1.3.7 *Effort–reward imbalance*

An individual puts effort into certain tasks for an equal and fair return – a reward (Smith et al. 2005). A reward, which originates from the basic social norm of 'reciprocity', can be distributed to employees by three means: money, esteem and career opportunities (De Jong et al., 2000). Although the degree of equality and fairness of the return can be subjective, it is largely governed by social norms. Insufficient financial, institutional or social rewards devalue the performance of employees (Maslach, Schaufeli and Leiter 2001) and it can be stressful (Maslanka 1996; Niedhammer et al. 2006; Siefert, Jayaratne and Chess 1991). Fair rewards satisfy construction personnel's basic needs to survive and, more importantly, act as recognition of their performance and value to the organisation. Previous studies have confirmed the significant impact of effort–reward imbalance on the stress of estimators and construction workers (Leung, Sham and Chan 2007; Leung, Chan and Yuen 2010).

Construction workers are normally placed at the lowest level of an organisation's structure and are forced to exert considerable physical effort to undertake their tasks in a poor environment. In view of these unfavourable conditions, construction workers are often reported to have unfair rewards in terms of salary, working hours, respect, labour welfare, career development and so on (Stattin and Jarvholm 2005; The Standard 2007). An effort–reward imbalance has a direct impact on stress levels and demotivates employees (Cordes and Dougherty 1993).

4.1.3.8 *Provision of adequate safety equipment*

There are legislations (e.g., the Employer's Liability Act 1969 in the UK) governing employers to provide suitable safety equipment to construction workers (Cooke 2009). In cases of construction workers being injured due to inadequate personal protective equipment provided, the employers would be liable for the negligence. However, construction personnel are unfortunately often found to be working with inadequate safety equipment (Hinze 1988). Provision of safety equipment is essential to preventing injury among workers, as is having the appropriate types of equipment and the maintenance of the equipment (Sunindijo and Zou 2013). Working with insufficient or inappropriate safety equipment certainly induces stress among rational construction workers, since they worry about their safety in the crisis-ridden site environments every day.

Previous research has found scientific support for a direct and significant impact of inadequate safety equipment on the emotional stress levels of construction workers (Leung, Chan and Yuen 2010; Leung, Chan and Yu 2011). The safety budget should always be the top priority in a project, even in competitive bidding (Zou, Zhang and Wang 2007). Adequate safety equipment can smoothen the work process on site and demonstrate that the organisation cares about its construction workers, which is essential for developing a sense of belongingness among construction workers. This is especially essential for today's construction industry, in which aging workers and understaffing are becoming more and more common. A caring organisational climate can help alleviate the emotional stress of construction workers (MacDavitt, Chou and Stone 2007).

4.1.4   Organisational Stressors

4.1.4.1 *Organisational centralisation*

Centralisation concerns the decision-making authority of personnel in an organisation (Francesco and Gold 2005). The degree of organisational centralisation can be identified by the hierarchy of authority and the degree of members' participation in the decision making of an organisation (Andrews, George and Richard 2009; Glisson and Martin 1980). In a centralised organisation, only a small number of staff members from top management levels are allowed to participate in the decision-making process. For a decentralised organisation, a larger number of staff and those from middle or lower management are also involved in the decision-making process (Francesco and Gold 2005).

Previous studies have found that centralisation promotes cost control through standardisation, while decentralisation promotes organisation innovation by allowing flexibility for new and creative ideas in the production or management processes (Miller 1987). Organisational centralisation may influence the stress level of construction personnel because it determines the authority they are allowed in the organisation.

4.1.4.2 *Organisational formalisation*

Organisational formalisation concerns the degree to which an organisation is person- versus task-oriented (Joiner 2001). Highly formalised organisations tend to be task-oriented (i.e., rule of law), wherein employees' behaviours are strictly governed by rules, policies and procedures specified in written guidelines and regulations (Francesco and Gold 2005). In contrast, an organisation with low formalisation tends to be person-oriented (i.e., rule of man). Formalisation is important for maintaining the normal operations and cost efficiency of an organisation through standardised procedures and systemised information (John and Martin 1984; Miller 1987; Segars, Grover and Teng 1998). However, too much formalisation can limit the innovation of an organisation – a key performance indicator in the construction sector nowadays. This is because employees have less flexibility to acquire and utilise new knowledge and ideas in their daily production or management process (Kohli and Jaworski 1990; Miller 1987; Mintzberg 1979), which may cause stress.

In sum, organisational structure has a direct impact on the degree of bureaucracy in a company and the power allocated to its employees, both of which are critical stressors (Zaleznik, Kets de Vries and Howard 1977). Due to the dynamic nature of construction projects, an adaptive organisational structure facilitating a work climate with adequate autonomy and flexibility is essential to the health and performance of construction personnel. In fact, organisational structure and its resulting climate have been found to be a stressor for construction personnel (Sutherland and Davidson 1993; Lossemore and Waters 2004). In addition, a deficient organisational structure will have a negative impact on employees' work attitudes, behaviours, motivation, commitment, productivity and satisfaction (Child, Faulkner and Tallman 2005; Vigoda 2000). It definitely causes stress among employees and induces real costs for the organisation due to worsened individual performance and, more importantly, project performance (Kasl 1992; Rogers 1983).

4.1.4.3 *Organisational support*

According to organisational support theory, employees are concerned about the degree to which the organisation values their contributions and cares about their well-being. This is important to their socio-emotional needs and whether they intend to perform highly in the future or not (e.g., performance improvement and withdrawal behaviours; Eisenberger et al. 1986; Rhoades and Eisenberger 2002). The support provided by an organisation can generally be categorised into assignment-related support (e.g., manpower, material resources and technical support) and non-assignment-related support (e.g., financial support and career development support; Kraimer and Wayne 2004; Takeuchi, Shay and Li 2009).

In fact, organisational support has been found to contribute to individual socio-emotional problems, including individual work-related stress (Lazarus and Folkman 1984), psychological and psychosomatic reactions (George et al. 1993), burnout (Cropanzano et al. 1997) and emotional exhaustion (Eriksson et al. 2009). Moreover, organisational support can moderate the relationship between individual stress and job performance (AbuAlRub 2004; Ismail et al. 2010). Specifically, if employees perceive higher organisational support, their job performance under stressful situations is less likely to be negatively affected.

4.1.5   Physical Stressors

4.1.5.1 *Work environment in the office*

'Physical stressors' simply refers to the poor environment around the individual (Driskell and Salas 1991; Quick and Quick 1997), which can mainly arise from the working environment and home environment. For construction personnel, the working environment involves both the head office and on-site environment. The office environment is essential to the health and safety of employees, including single-room versus open-plan offices (determining the degree of noise), the presence, location and size of partitions (determining the degree of privacy), windows (determining the degree of ventilation and lighting) and air-conditioning (determining the degree of temperature and ventilation). In fact, the office environment affects both physical and psychological health and further influences work performance and social behaviours (Lewy et al. 1982). For instance, employees will suffer from excessive noise if they are assigned a seat near a plant room without any sound insulation partition. Excessive and continuous exposure to noise causes various physical stress symptoms (e.g., high blood pressure) and negatively affects individual psychosocial relationships and work performance (e.g., disrupted attention and memory on complex tasks; Driskell and Salas 1996).

4.1.5.2 *Work environment on site*

Unlike professionals in other industries, construction personnel are often required to work between head office and sites (Leung et al. 2005; Leung, Zhang and Skitmore 2008). Therefore, they are not only affected by the office environment, but also the construction site environment. In fact, the extensive literature has confirmed the impact of a poor on-site environment – such as an extreme temperature, excessive noise and vibration and overcrowdedness due to a large number of workers, equipment or materials – on the stress of construction personnel (CIOB 2006; Djebarni 1996; Leung et al. 2005; Leung, Skitmore and Chan 2007; Leung, Chan and Olomolaiye 2008; Leung, Zhang and Skitmore 2008; Leung, Chan and Yu 2012). For instance, the temperature on site is often extreme because work sites are uncovered. Prolonged exposure to high temperature results in impaired and disrupted attention, memory, judgment and piloting skills. Meanwhile, prolonged exposure to low temperature induces slow responses, loss of manipulative ability (hands) and effects on cognitive tasks (Driskell and Salas 1996; Yip and Rowlinson 2009).

The impact of a poor on-site environment is particularly significant for construction workers because they need to stay on site for the entire working day. In addition, employers are responsible for providing safe working environment to construction personnel and suitable safety policy should be in place so that construction personnel can be instructed what and how to work safely (Cooke 2009). Inadequate safety measures (e.g., lack of protective measures on site) cause stress and feelings of insecurity for construction workers.

Prolonged periods of working in adverse physical environments induces stress (Vischer 2007), affects their safety and their ability to perform tasks (Driskell and Salas 1991; Goldenhar, Williams and Swanson 2003) and, more importantly, causes higher injury rates among construction workers (Leung, Chan and Yuen 2010).

4.1.5.3 Home environment

In addition to the work environment, the home environment is also critical to people's stress level. Compared to an impoverished home environment, an enriched, comfortable environment enhances various cognitive and sensory functions and relieves the emotional disturbances induced by stress (Schloesser et al. 2010). Poor home environments have thus been found to have a significant impact on both work and emotional stress of construction personnel (Leung et al. 2008). In fact, one study found that a poorly functioning home or non-job environment affects people's emotional state and their performance at work (Organ 1979). Hence, in the quest to improve performance by properly managing construction employees' stress levels, their home environment stressors cannot be ignored.

In fact, as mentioned in the hierarchy of needs theory, the home environment is one of the important basic physiological needs of an individual (Maslow 1954). Only after the basic needs are satisfied will an individual proceed to acquire higher-order needs, including needs of safety, affiliation, esteem and self-actualisation. To a certain extent, housing benefits help fulfil the physical shelter needs of an individual and they are essential in motivating employees to put forth effort in their jobs and at the same time, releasing their stress in fulfilling the personal or familial housing needs.

4.2   Development of a Conceptual Model of Stressors and Stress

Based on the extensive literature on stress, it can be said that there are 21 stressors in five main stressor categories that cause stress among construction personnel: personal, interpersonal, task, organisational and physical stressors. Figure 4.1 illustrates the overall five-stressors model.

In fact, these five main stressor categories are not separate mechanisms. As shown in Figure 4.1, there is overlap between various types of stressors, implying that they interact with each other. For instance, a study of construction estimators found that work overload triggers work–home conflict (Leung et al., 2005). On the other hand, another study found that poor office environments worsen interpersonal relationships, further inducing role ambiguity and role conflict among construction project managers (Leung et al., 2009). These findings indicate that different types of stressors actually have close interactions with each other and that the impact on stress can both be direct and indirect. Therefore, stressor management should not focus on only one type of stressor, but consider the comprehensive influence of different factors.

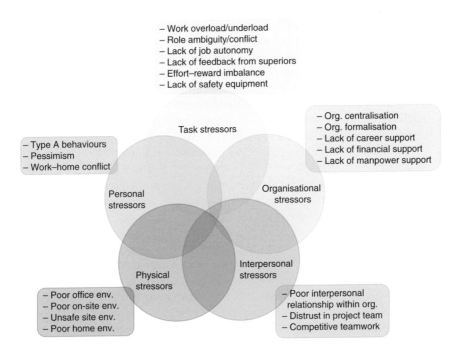

Figure 4.1   Conceptual model of stressors influencing construction personnel.

4.3   Research Results on Stressors and Different Construction Personnel

Following the review of the literature on the various types of stressors in the past decades, this section explores and outlines scientific evidence for the experiences of various types of stressors among different types of construction personnel (including both professionals and manual workers). To achieve this aim, this section reports a study using both quantitative methods and case studies.

4.3.1   Statistical Studies of Stressors of Construction Personnel

To measure the stress levels of construction personnel, 60 items were adopted from our previous studies on the different types of personal, interpersonal, task, organisational and physical stressors (adopted/developed and validated by Leung et al. 2005; Leung, Skitmore and Chan 2007; Leung, Chan and Olomolaiye 2008; Leung, Zhang and Skitmore 2008; Leung, Chan and Yu 2011; Chan, Leung, Chan and Yu 2012 in the construction context). A total of 702 construction personnel were asked to rate their agreement with various stressor items using a 7-point Likert-type scale, ranging from 1 (*strongly disagree*) to 7 (*strongly agree*). Twenty-one stressors were analysed within five categories (i.e., personal, interpersonal, task, organisational and physical). Data analyses were conducted by SPSS 20.0.

Due to the different sample sizes, mean scores and standard deviations were calculated for all of the stressors. Among all the stressors, *inadequate safety equipment, poor site environment and organisational formalisation* got the highest mean scores with values of 4.838, 4.679 and 4.577 in the task, physical and organisational categories, respectively. The means for all stressors are shown in Table 4.1 and Figure 4.2. The standard deviations for all the stressors were not large, ranging between 0.83 and 1.39.

Constructions sites have many hazards, such as working at height, sharp edges, deep foundation pits and electrical equipment. People working on construction sites often face many dangers. For instance, they may be hurt by machines, cut by sharp edges, injured by nails, tripped by holes or hit by falling construction materials (e.g., concrete blocks and bricks). Rational construction workers without adequate safety equipment support will worry about their safety and have little sense of security while working on site and thus they cannot fully concentrate on their work. For example, a scaffolding operator would be in severe danger if there is inadequate personal protective equipment like safety helmets, belts and lifelines. Due to the serious consequences of the inadequate safety equipment, it is no wonder that it was one of the top three highly scored stressors out of 21, which fits with the results of previous studies (Leung et al., 2010, 2011).

The environment of construction sites can be poor. The air quality on temporary sites is often poor due to the dust and particulate pollution from construction and plant operations. Construction personnel need to work exposed to the weather in poor conditions such as extreme temperature, excessive noise and vibration. Staying in such poor site conditions affects them not just physically (e.g., sun burns and heatstroke) but also

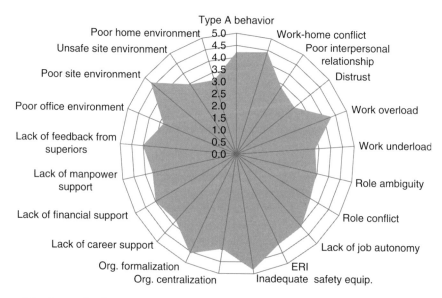

Figure 4.2 Stressor levels experienced by construction personnel. Note: ERI = effort–reward imbalance; Org. = organisation; Equip. = equipment.

Table 4.1   Stressors of construction personnel.

Stressors	Sample size	Mean score	SD	Alpha
Personal				
1. Type A behaviour	230	4.217	1.026	0.636
+ I am an achievement-oriented person who has the need to win				
+ I enjoy competition and feel I always have to win				
+ People sometimes say that I have a quick temper				
2. Work–home conflict	324	4.423	1.118	0.626
+ My family/marriage relationships suffer because of my work				
+ Because of my work, I am not as a good parent/partner as I would like to be				
+ My job leaves me plenty of time to spend with my friends				
Interpersonal				
3. Poor interpersonal relationship	411	3.367	0.828	0.629
− I have a good relationship with my supervisor				
− My colleagues are trustworthy and friendly				
+ Relationships with others are complicated				
4. Distrust	68	3.140	1.178	0.769
+ There often seems to be a lack of trust between me and my subordinates				
+ I seldom delegate tasks because others cannot complete the tasks as well as I can				
+ My colleagues often compete with one another rather than cooperate with a feeling of team spirit				
Task				
5. Work overload	497	4.315	1.348	0.670
+ My job uses up most of my time to relax				
+ There is constant pressure to work all the time				
+ I have a lot of responsibility in my job				
6. Work underload	140	3.343	1.053	0.661
+ I feel my skills and abilities are not being used well				
+ I frequently find my job boring and repetitive				
+ I am given very little responsibility				
7. Role ambiguity	620	3.472	1.271	0.881
+ My job responsibilities are generally vague, unclear and inconsistent				
+ My goals and objectives are intangible and not clearly spelled out				
+ Explanations of what has to be done are often unclear				
8. Role conflict	140	3.621	0.975	0.598
+ My beliefs often conflict with those of the organisation				
+ I am often caught by conflicting demands between different parties				
+ Things I do are often accepted by one person but not another				
9. Lack of job autonomy	139	4.055	0.750	0.622
+ I have to refer matters upwards when I can really adequately deal with them myself				

(Continued)

Table 4.1 (Continued)

Stressors	Sample size	Mean score	SD	Alpha
+ My supervisor often deals with me in an autocratic and overly demanding manner + I was given insufficient authority to do my job properly				
10. Effort–reward imbalance + I find the reward I get is relatively low when compared to the external market + I often feel that the organisation treats us unfairly + I find the reward I get is not balanced with the effort I put in	276	4.118	1.390	0.892
11. Inadequate safety equipment – Safety equipment is in good condition – Safety equipment is regularly checked – Safety equipment is under regular maintenance	276	**4.838**	1.385	0.938
12. Lack of feedback from superiors – My immediate supervisor gives prompt attention to problems on my job – My supervisor gives me recognition when I do a good job	274	4.013	1.369	0.795
Organisational				
13. Organisational centralisation + Subordinates have to ask their superiors before they can do almost anything + Even small matters have to be referred to someone higher up for the final answer + Any decision I make has to have my boss's approval	135	3.965	1.283	0.814
14. Organisational formalisation + A 'Rule & Procedures' manual exists and is readily available within the department + I have to carry out the work in complicated work procedures + The company emphasises paper work and documentation	137	**4.577**	1.069	0.643
15. Lack of career support – My company provides me with stable career and promotion opportunities – My company takes an interest in my career – The company provides a mentoring scheme for me	134	3.801	1.107	0.779
16. Lack of financial support – My company has taken care of me financially – The financial incentives and allowances provided by my company are good – I have received generous financial support from my company	137	3.992	1.190	0.868
17. Lack of manpower support + I do not have sufficient supporting staff to assist my work + I have to consistently 'borrow' staff from another team + There is insufficient supporting staff to finish my work	68	3.677	1.255	0.842
Physical				
18. Poor office environment + My office is too crowded + My office is abnormally noisy + The temperature of my work place is too extreme + The lighting in the office is too dim	287	3.345	1.118	0.782

Table 4.1   *(Continued)*

Stressors	Sample size	Mean score	SD	Alpha
19. Poor site environment	274	**4.679**	1.289	0.875
+ The air quality of the site is poor				
+ The construction site is noisy				
20. Unsafe site environment	276	3.465	1.048	0.639
+ Equipment is placed in an unorganised way				
− Lighting is sufficient				
− Provision of indicators is sufficient				
21. Poor home environment	81	3.191	1.355	0.802
+ My home environment needs adjustment				
+ I am not satisfied with my home environment				

Note: **Bolded** items - highest mean scores obtained +/-denotes items with aligned or opposite direction with the indicating stressor (for '-' items, values were reversed in the data analysis process).

psychologically. Hence, construction personnel rated poor on-site environment as the second highest stressor in the study.

Construction firms often involve professionals from multiple disciplines (e.g., project managers, engineers and manual labourers) in different levels (senior or junior) for various tasks (Planning, designing, and construction). In order to cooperate with multidisciplinary professionals in various departments or teams in an organization, it is necessary to establish standardised guidelines. Formalised organisations tend to establish systematic guidelines and regulations for standardised working procedures, information-processing systems and so on. Employees are governed by guidelines and regulations strictly while carrying out their work (Francesco and Gold 2005). Hence, these professionals may find that their work is too rigid, with too little flexibility to perform their tasks. Prolonged work under such conditions will cause them stress. For example, organisational regulations require a building service engineer to make enquires or respond to enquiries from other project participants in formal ways, such as sending formal emails, no matter how minor the issues are. Extremely formalised work procedures induce a great deal of work stress on the engineer (increasing quantitative workload) and delay the construction process. Hence, although formalisation is important for organisations to maintain and manage normal operations, it can be a source of stress to employees, particularly for construction personnel, who have to deal with various dynamic tasks every day.

4.3.1.1 *Comparison of stressors experienced by various construction personnel groups*

In order to investigate the impact of construction disciplines on levels of stressors, one-way between-groups analyses of variance were conducted. The results revealed no significant difference in the mean value of stressors

Table 4.2 One-way between-groups ANOVA for stressors of construction personnel in different age groups.

Stressors	Age Group	Mean	SD	F (ANOVA)	Sig. (ANOVA)	Sig. (Levene)	Group	Mean Diff.	S.D.	Sig.
							Post-hoc test for stressors with significant diff. scores			
1. Type A behaviour	≤ 29	4.279	1.130	0.198	0.898	0.588	–	–	–	–
	30–39	4.288	1.013							
	40–49	4.251	1.279							
	≥ 50	4.120	1.191							
2. Work–home conflict	≤ 29	4.356	0.886	1.830	0.140	0.019	–	–	–	–
	30–39	4.419	0.824							
	40–49	4.543	0.746							
	≥ 50	4.450	0.887							
3. Poor interpersonal relationship	≤ 29	3.374	0.718	1.091	0.352	0.113	–	–	–	–
	30–39	3.272	0.716							
	40–49	3.277	0.593							
	≥ 50	3.241	0.735							
4. Distrust	≤ 29	3.153	0.273	**2.861**	**0.036**	**0.011**	–	–	–	–
	30–39	**3.199**	**0.443**				40–49	0.097	0.036	0.033
	40–49	3.102	0.399							
	≥ 50	3.106	0.228							
5. Work overload	≤ 29	4.131	1.192	**2.674**	**0.045**	**0.254**	≤ 29	0.323	0.115	0.026
	30–39	4.295	1.128							
	40–49	**4.454**	**1.116**							
	≥ 50	4.310	1.069							
6. Work underload	**≤ 29**	**3.388**	**0.548**	**5.688**	**0.001**	**0.013**	40–49	0.188	0.050	0.001
	30–39	3.273	0.463				≥ 50	0.196	0.060	0.006
	40–49	3.200	0.503							
	≥ 50	3.192	0.404							
7. Role ambiguity	≤ 29	3.608	1.196	1.749	0.156	0.067	–	–	–	–
	30–39	3.448	1.127							
	40–49	3.367	1.194							
	≥ 50	3.319	1.334							

		Mean	SD							
8. Role conflict	≤ 29	3.753	0.612	0.311	0.817	0.565	–	–	–	–
	30–39	3.706	0.747							
	40–49	3.719	0.712							
	≥ 50	3.673	0.721							
9. Lack of job autonomy	**≤ 29**	**4.185**	**0.830**	**2.740**	**0.042**	**0.001**	**≥ 50**	**0.255**	**0.092**	**0.028**
	30–39	4.037	0.756							
	40–49	4.044	0.626							
	≥ 50	3.931	0.847							
10. Effort–reward imbalance	**≤ 29**	**4.294**	**1.037**	**3.564**	**0.014**	**0.000**	**30–39**	**0.283**	**0.094**	**0.014**
	30–39	4.011	0.865							
	40–49	4.124	0.764							
	≥ 50	4.028	0.810							
11. Inadequate safety equipment	≤ 29	4.878	0.901	1.904	0.128	0.106	–	–	–	–
	30–39	4.933	0.838							
	40–49	4.739	0.771							
	≥ 50	4.856	1.029							
12. Lack of feedback from superiors	≤ 29	4.250	1.354	1.352	0.258	0.251	–	–	–	–
	30–39	3.820	1.541							
	40–49	3.971	1.213							
	≥ 50	3.921	1.378							
13. Organisational centralisation	≤ 29	3.950	0.434	2.368	0.070	0.005	–	–	–	–
	30–39	4.039	0.602							
	40–49	3.899	0.599							
	≥ 50	4.001	0.559							
14. Organisational formalisation	≤ 29	4.573	0.294	0.075	0.974	0.009	–	–	–	–
	30–39	4.567	0.542							
	40–49	4.588	0.505							
	≥ 50	4.576	0.492							
15. Lack of career support	≤ 29	3.749	0.355	0.423	0.737	0.246	–	–	–	–
	30–39	3.792	0.477							
	40–49	3.783	0.521							
	≥ 50	3.809	0.504							

(Continued)

Table 4.2 *(Continued)*

Stressors	Age Group	Mean	SD	F (ANOVA)	Sig. (ANOVA)	Sig. (Levene)	Group	Mean Diff.	S.D.	Sig.
		One-way between-groups ANOVA					Post-hoc test for stressors with significant diff. scores			
16. Lack of financial support	≤ 29	3.965	0.323	0.274	0.844	0.033	–	–	–	–
	30–39	4.015	0.539							
	40–49	3.986	0.614							
	≥ 50	3.985	0.457							
17. Lack of manpower support	≤ 29	3.670	0.326	0.141	0.935	0.017	–	–	–	–
	30–39	3.692	0.499							
	40–49	3.669	0.390							
	≥ 50	3.679	0.242							
18. Poor office environment	**≤ 29**	**3.496**	**0.663**	**3.731**	**0.011**	**0.442**	**40–49**	**0.224**	**0.072**	**0.011**
	30–39	3.338	0.761				**≥ 50**	**0.230**	**0.087**	**0.043**
	40–49	3.272	0.751							
	≥ 50	3.266	0.609							
19. Poor site environment	**≤ 29**	**4.770**	**0.830**	**4.074**	**0.007**	**0.000**	**≥ 50**	**0.330**	**0.099**	**0.005**
	30–39	**4.712**	**0.766**				**≥ 50**	**0.271**	**0.097**	**0.026**
	40–49	**4.680**	**0.736**				**≥ 50**	**0.240**	**0.092**	**0.046**
	≥ 50	4.440	0.962							
20. Unsafe site environment	≤ 29	3.505	0.701	0.774	0.509	0.006	–	–	–	–
	30–39	3.404	0.640							
	40–49	3.482	0.596							
	≥ 50	3.456	0.730							
21. Poor home environment	≤ 29	3.207	0.155	1.109	0.345	0.000	–	–	–	–
	30–39	3.185	0.484							
	40–49	3.164	0.551							
	≥ 50	3.255	0.444							

Note: **Bolded** figures - significant between-group differences

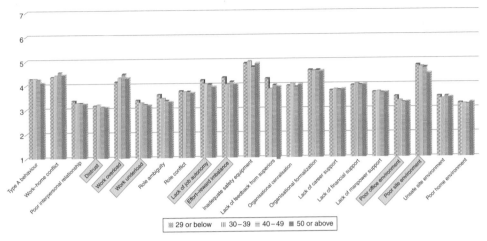

Figure 4.3 Stressor levels of construction personnel in different age groups Note: Highlighted items– Significant between-group differences in the one-way ANOVA (Table 4.2) ERI = Effort–reward imbalance; Org. = Organisation; Equip. = Equipment.

between different construction personnel and gender groups, although there were significant differences in stressors between different age groups. As shown in Table 4.2 and Figure 4.3, survey respondents were divided into four groups (Group 1: 29 years and younger; Group 2: 30 to 39 years; Group 3: 40 to 49 years; Group 4: 50 years and older). Out of the total 702 participants, 23.2% were 29 years or younger, 25.9% were 30–39, 34.3% were 40–49 and 16.1% were 50 and older. Among all the 21 stressors, seven had statistically significant differences at the $p < 0.05$ level for the four age groups: distrust ($F = 2.861$, $p < 0.05$), work overload ($F = 2.674$, $p < 0.05$), work underload ($F = 5.688$, $p < 0.01$), lack of job autonomy ($F = 2.740$, $p < 0.05$), effort–reward imbalance ($F = 3.564$, $p < 0.05$), poor office environment ($F = 3.731$, $p < 0.05$), and poor on-site environment ($F = 4.074$, vp < 0.05). To further compare the stressors with significant differences between the four age groups, Tukey's HSD tests was conducted for the seven stressors.

Distrust is very important for construction personnel and it differed significantly between age groups. Among the four age groups, distrust for was highest for construction personnel aged 30–39. The results of the post hoc test showed that the mean score for this group of construction personnel ($M = 3.199$, $SD = 0.443$) was significantly different from construction personnel aged 40–49 ($M = 3.102$, $SD = 0.399$), with a mean difference of 0.097 ($p < 0.05$). There was no significant difference between the other two groups. Construction personnel aged 30–39 have amassed a certain amount of working experience, in which they are usually leaders of a group of younger construction personnel. Since these younger construction employees often have comparatively limited management experience, construction personnel in the 30–39 age group may prefer to perform the tasks by themselves, since they are often driven by tight project schedules. On the other hand, construction personnel aged 40–49 have often gained much richer experiences and their immediate subordinates maybe be more experienced (usually those in the 30–39 age group). Hence, construction personnel aged 40–49 rated distrust as less of stressor.

The level of *work overload* of construction personnel aged 40–49 ($M = 4.454$, $SD = 1.116$) was higher than that experienced by their counterparts aged 29 or below ($M = 4.131$, $SD = 1.192$), with a mean difference of 0.323 ($p < 0.05$). There was no significant difference between the other groups. Construction personnel aged 40–49 have usually amassed sufficient experience. They often play the role of supervisor in project teams and manage crises when they occur. Hence, this group of construction personnel often faces job tasks, which are more complex and challenging. Thus they have a higher likelihood of suffering from overload on the job. However, since younger personnel are freshmen in their organisations, they are often assigned less important and less complicated tasks. In large-scale construction firms, these young personnel are often assigned mentors whom they can depend on whenever they face difficulties on the job. Therefore, the perceived work overload of construction personnel aged 29 or below is lower than that of the other three groups, especially those aged 40–49.

The *work underload* stressor was the highest for construction personnel aged 29 or below ($M = 3.39$, $SD = 0.548$), which was significantly different from personnel aged 40–49 ($M = 3.200$, $SD = 0.503$) and 50 or below ($M = 3.192$, $SD = 0.404$) respectively, with mean differences of 0.188 ($p < 0.01$) and 0.196 ($p < 0.01$) respectively. There were no significant differences between the other groups. Younger construction personnel (i.e., aged 29 or below) are usually assigned comparatively simple or repetitive tasks, such as drawings, measurement, payment assessment and programme preparing. Hence, they may feel their tasks are boring and not challenging and experience work underload. In contrast, construction personnel aged 40 and above have gained a lot of working experience and are normally responsible for complicated and important tasks in a project. Therefore, construction personnel aged 40 and above were found to experience a significantly lower level of work underload.

Similarly, construction personnel aged 29 or below were found to experience the highest level of *lack of job autonomy* ($M = 4.185$, $SD = 0.830$), which was significantly higher than personnel aged 50 or above ($M = 3.931$, $SD = 0.847$), with a mean difference of 0.255 ($p = 0.028$). Compared to construction personnel in other groups, those in the age range below 29 are rookies or even fresh graduates from universities and often lack of practical experience. Hence, they receive a higher degree of mentorship and instructions from their superiors and other experienced personnel rather than carrying out their tasks independently. Thus, it is easy for them to suffer from the lack of job autonomy in the beginning of their professional career. On the other hand, construction personnel aged 50 or above have richer experience and knowledge to perform their job with certain autonomy and they have the authority and professional qualifications to plan their own job and the jobs of others. Hence, construction personnel aged below 30 are more susceptible to lack of job autonomy than those aged 50 and above.

Effort–reward imbalance was rated the highest by construction personnel aged 29 and below (M=4.294, SD=1.037), which was significantly higher than those aged 30–39 (M=4.011, SD=0.865), with a mean difference of 0.283 (p<0.05). Young construction personnel have just started their careers and have limited work experience in the industry. Due to their junior status, they are more often impelled to work extra time than their seniors on assigned tasks in order to establish a solid platform for further development. Therefore, it was found that young construction personnel tend to consider their effort out of balance with the reward they receive. After amassing certain experience, construction personnel aged 30–39 have progressively established professionalism in their careers in terms of both qualifications and position. Their rewards, including monthly salary and benefits, are then increased accordingly. Thus, they experience less effort–reward imbalance.

It is also found that the youngest group of construction personnel (aged 29 or below) scored the highest on *poor office environment* (M=3.496, SD=0.663), which was significantly higher than personnel aged 40–49 (M=3.272, SD=0.751) and 50 and above (M=3.266, SD=0.609), with the mean differences of 0.224 (p<0.05) and 0.230 (p<0.05) respectively. As junior staff, young construction personnel may be assigned to sit and work with a group of junior members, in which the space is often open, crowded and noisy, compared with the work environment of senior members. Construction personnel aged 40 or above are often working at senior positions in an organisation, in which they are often assigned spaces with larger area and higher privacy.

Construction personnel aged 50 or above (M=4.440, SD=0.962) suffer significantly less from *poor site environment* than their counterparts aged 29 and below (M=4.770, SD=0.830), 30–39 (M=4.712, SD=0.766) and 40–49 (M=4.680, SD=0.736), with mean differences of 0.330 (p=0.005), 0.271 (p=0.026) and 0.240 (p=0.046) respectively. Since construction personnel aged 50 or above have amassed a lot of experience on site, it is no surprise that they are more tolerant of the poor on-site environment than the other three younger groups. Unlike the office environment, construction sites normally contain waste, such as soils and frameworks with exposed nails. Staying on the site is really dangerous and usually takes a lot of energy (Djebarni 1996, Leung et al. 2005; Leung, Skitmore and Chan 2007; Leung, Sham and Chan 2007; Leung, Chan and Olomolaiye 2008; Leung, Zhang and Skitmore 2008; Leung, Chan and Yu 2010b). Moreover, the on-site environment is often outdoors, where construction personnel are exposed to different types of weather, such as extreme temperature, bright sunlight, heavy rain and snow and strong wind. Hence, construction personnel, especially the young groups, generally rated the site environment as poor.

In sum, young construction personnel can easily suffer from work underload, work overload, lack of job autonomy, effort–reward imbalance and poor office/site environments, while those aged 30 to 39 have significantly higher distrust of their colleagues and team members. The oldest construction personnel are in the late stage of their career. They earn higher incomes and have less appetite for promotion compared with other groups. Therefore, they can enjoy their work life, including job autonomy, physical environment and work underload, while preparing for retirement.

4.3.1.2 *Correlations between stress and stressors*

To investigate the relationships between stressors and stress of construction personnel, Pearson correlation analysis was adopted in the study (see Table 4.3). The results revealed that two personal stressors (Type A behaviour and work–home conflict), one interpersonal stressor (poor interpersonal relationship), five task stressors (work overload, role ambiguity, role conflict, effort–reward imbalance and lack of feedback from superior), one organizational stressor (lack of manpower support), and two physical stressors (poor office environment and unsafe site environment) were significantly correlated to all three types of stress (work, physical and emotional). In contrast, two organisational stressors (organisational formalisation and lack of financial support) have no significant relationship with stress or with other stressors. Therefore, the two organisational stressors are not presented in Table 4.3.

1. *Stress and personal / interpersonal stressors.* The results showed that three (inter-) personal stressors (Type A behaviour, work–home conflict and poor interpersonal relationships) elevate all three types of stress, while distrust only associates with emotional stress. According to previous studies, Type A people report more stress and stress-related illness than Type B people do (Caplan and Jones 1975; French and Caplan 1972; Friedman and Rosenman 1974; Leung, Sham and Chan 2007; Leung, Chan and Olomolaiye 2008). Construction personnel with *Type A behaviours* tend to strive for achievement, behave competitively, act hostile and spend a lot time on their work, which causes them much stress at work. The prolonged time and high degree of effort and attention put into work also tires them out both emotionally and physically.

    Meanwhile, stress is also significantly related to construction personnel's family and interpersonal relationships. A good balance between *work and home* is a key to maintaining individual well-being (Lingard and Francis 2006). Excessive work demands interfere with the quality of family life of construction personnel, driven by the responsibility and commitment towards both sides, both emotionally and physically. In fact, Type A individuals are more dominant in *interpersonal relationships* than Type B people and impatient in competitive situations (Yarnold and Grimm 1982). Consistent with previous findings, poor interpersonal relationships with colleagues, including subordinates and supervisors, increase construction professionals' stress levels (Leung 2004). Good relationships with workgroups are important for the complicated tasks of construction personnel who are involved in the dynamic construction projects with multiple stakeholders (Djebarni 1996). Influenced by poor workgroup relationships, construction personnel may also be frustrated, unhappy and worried (i.e., emotional

Table 4.3 Correlations of the stressors of construction personnel.

Stressors (Ss)	Stress			Personal		Interpersonal		Task								Organisational			Physical		
	WS	ES	PS	Ss1	Ss2	Ss3	Ss4	Ss5	Ss6	Ss7	Ss8	Ss9	Ss10	Ss11	Ss12	Ss13	Ss14	Ss15	Ss16	Ss17	Ss18
1. Type A behaviour	.089*	.222**	.260**	1.000	–	–	–	–	–	–	–	–	–	–	–	–	–	–	–	–	–
2. Work–home conflict	.153**	.326**	.239**	.268**	1.000	–	–	–	–	–	–	–	–	–	–	–	–	–	–	–	–
3. Poor interpersonal relationship	.190**	.314**	.210**	.139*	.111**	1.000	–	–	–	–	–	–	–	–	–	–	–	–	–	–	–
4. Distrust	–	.140**	–	–	–	.089*	1.000	–	–	–	–	–	–	–	–	–	–	–	–	–	–
5. Work overload	.307**	.401**	.257**	.213**	.310**	.240**	–	1.000	–	–	–	–	–	–	–	–	–	–	–	–	–
6. Work underload	–	–	.107**	–	–	–	–	–	1.000	–	–	–	–	–	–	–	–	–	–	–	–
7. Role ambiguity	.104**	.292**	.245**	.211**	.208**	.126**	–	.221**	.235**	1.000	–	–	–	–	–	–	–	–	–	–	–
8. Role conflict	.156**	.312**	.231**	–	.316**	.260**	.142**	.242**	.320**	.326**	1.000	–	–	–	–	–	–	–	–	–	–
9. Lack of job autonomy	–	–	–	–	.220**	.287**	–	.240**	–	.299**	.223**	1.000	–	–	–	–	–	–	–	–	–
10. Effort–reward imbalance	.130**	.368**	.259**	.293**	.292**	.423**	–	.348**	–	.365**	.253**	.530**	1.000	–	–	–	–	–	–	–	–
11. Inadequate safety equipment	–	.205**	.119**	–	–	.396**	–	.127**	–	.202**	.187**	.110**	.276**	1.000	–	–	–	–	–	–	–
12. Lack of feedback from superiors	.097*	.262**	.252**	–	.228**	.298**	–	.291**	–	.332**	.251**	.443**	.405**	.130**	1.000	–	–	–	–	–	–
13. Organisational centralisation	–	.142**	–	–	–	–	–	–	–	.162**	–	–	–	–	–	1.000	–	–	–	–	–
14. Lack of manpower support	.116**	.193**	.106**	–	–	–	.341**	–	–	–	.127**	.076**	–	–	–	–	1.000	–	–	–	–

(Continued)

Table 4.3 (Continued)

Stressors (Ss)	Stress			Personal		Interpersonal		Task								Organisational			Physical		
	WS	ES	PS	Ss1	Ss2	Ss3	Ss4	Ss5	Ss6	Ss7	Ss8	Ss9	Ss10	Ss11	Ss12	Ss13	Ss14	Ss15	Ss16	Ss17	Ss18
15. Lack of career support	–	.101**	–	–	–	–	–	–	–	.123**	–	–	–	–	–	.257**	–	1.000	–	–	–
16. Poor office environment	.114**	.145**	.202**	–	–	.124**	–	–	.312**	.160**	.250**	–	–	–	–	–	.112**	–	1.000	–	–
17. Poor site environment	–	.132**	.154**	.164*	.185***	.124**	–	–	–	.118**	.103**	.244**	.223**	.147**	.306**	–	–	–	–	1.000	–
18. Unsafe site environment	.121**	.230**	.139**	–	.096**	.387**	–	.179**	–	.244**	.134**	.168**	.334**	.533**	.183**	–	–	–	–	.198**	1.000
19. Poor home environment	.084*	–	–	–	–	–	–	–	–	–	.081*	–	–	–	–	–	–	–	.185**	–	–

Note:

WS - work stress; ES - emotional stress; PS - physical stress

*correlation is significant at the 0.05 level (two-tailed) and

**correlation is significant at the 0.01 level (two-tailed).

Only variables with significant correlation(s) found are shown in this table.

Stressor in bold is significantly related to all three types of stress. *Stressor* in italic is significantly related to one or two types of stresses.

'_' denotes non-significant, duplicated or unavailable coefficients.

stress) due to the perception of personal inability to fulfil the contextual job requirements. Furthermore, faced with this stressor, construction personnel need to handle it and overcome it and its consequences, which are physically demanding and cause them physical stress. Lastly, since distrust, unlike explicitly poor interpersonal relationship, is implicit. Thus, its influence on people's work and cooperation may not be as direct as poor interpersonal relationships. Hence, it is only related to one's emotional state.

2. *Stress and task stressors.* Among the seven task stressors, work overload, role ambiguity, role conflict, effort–reward imbalance and lack of feedback from superiors were found to be significantly related to all three types of stress, while work underload was only correlated with physical stress and inadequate safety equipment was only related to emotional and physical stress. Although lack of job autonomy was not related to any kind of stress, it was significantly related to other stressors, which can in turn cause stress indirectly. Construction projects are dynamic and full of uncertainties. Construction personnel are often stressed by various urgent and complicated situations. When these *workloads* outweigh construction personnel's ability to cope, work stress can result. Influenced by these work demands in the long term, construction personnel may become emotionally fatigued and frustrated, resulting in emotional stress. In addition, the demanding job tasks often lead construction personnel to work overtime (Love, Haynes and Irani 2001). The long working hours affect their personal life and induce physical stress. On the other hand, construction personnel may suffer from *work underload*, which does not necessarily mean that they have too few tasks to do, but that their job tasks are too boring for them to fully utilise their ability (e.g., repetitive and tedious estimation tasks). Repetitive paper work causes physical problems for estimators, such as back pain. On the other hand, the fact that construction projects are often dynamic and involve multiple stakeholders is a key driver of poor *role ambiguity* and *role conflict* (Kahn et al. 1964). Suffering from poor role congruence, the boundary of work between project team members is not clear. This often leads to conflict in work cooperation, resulting in work stress and further affecting employees' emotional and physical well-being.

On the other hand, *effort–reward imbalance* is important for maintaining the well-being of individuals. In addition to achieving the physiological and safety needs of employees, a sense of effort–reward balance is important for satisfying other higher levels of needs, such as esteem and self-actualisation. It is a way for construction personnel to realise their value to the organisation. Effort–reward imbalance induces disappointment and frustration (De Jong et al. 2000) and it may cause emotional stress. Meanwhile, it discourages people from putting effort into their work tasks, which lowers their ability at work and then induces work and physical stress. Feedback from superiors offers guidance and recognition for one's accomplishments and is essential to the success of job tasks. Absence of feedback from superiors can induce depression, anxiety and unhappiness for construction personnel and in the long term can cause them to experience physical stress, which will lower their performance, thereby further resulting in work stress.

Furthermore, although organisations are responsible for providing the *safety equipment* necessary to protect construction personnel from harm, they may fail to do so because of tight project budgets or negligence (Hinze 1988; Zou, Zhang and Wang 2007). In these situations, construction personnel need to work on danger-ridden sites for tasks such as site inspection (construction professionals) and the construction itself (construction workers). Working with inappropriate safety equipment raises the likelihood that construction personnel will suffer from injuries (Leung, Chan and Yu 2010), which may cause them to be worried about their personal safety, leading to emotional and physical stress. In addition, construction personnel who work without appropriate safety equipment are more likely to suffer from *physical stress* problems (e.g., headaches arising from excessive noise in the case of a lack of ear plugs).

3. *Stress and organisational stressors.* Among the four organisational stressors, lack of manpower support were found to be associated with all three types of stress, while organisational centralisation and lack of career support were only correlated with emotional stress. An organisational culture and support that encourage *provision of adequate manpower resources* are essential (Ibem et al. 2011; Maslach, Schaufeli and Leiter 2001). Manpower support is critical for the often dynamic and urgent tasks of construction work. Failing to provide adequate manpower will cause both qualitative and quantitative difficulties in work tasks, resulting in work and physical stress, which will in turn affect construction personnel's emotional states.

    Under a *centralised organisational structure*, employees are working under a high degree of hierarchy of authority and low degree of participation in decision making (Andrews, George and Richard 2009; Carter and Cullen 1984; Glisson and Martin 1980), which means that their job authority is limited. Lack of job authority is found to predict depression, low self-esteem and job dissatisfaction (Beehr 1995; Leung, Sham and Chan 2007), which may result in emotional stress.

    Career support is an essential factor for retaining employees in the dynamic construction market (Lim and Ling 2012). Although it does not have a direct impact on work tasks, inadequate career support still causes worries and anxiety among construction personnel (Health and Safety Executive 2007), resulting in emotional stress. A comprehensive professional or career development scheme reduces emotional stress among construction employees and helps lower turnover rates.

4. *Stress and physical stressors.* In this study, poor office environments were found to induce all three types of stress. Poor site environments were significantly related to emotional and physical stress only, while poor home environment was only correlated with work stress. In fact, different individuals have different tolerance levels for adverse conditions (Gmelch 1982). However, the circumstances in which some tasks are carried out can create stress for most individuals and it is in those

situations that construction personnel are considered to have a poor work environment. When talking about the 'work environment', people usually refer to the *office environment* and such elements as the temperature, the lighting, the noise, the staff density and the degree of privacy. These can affect people's work efficiency, emotions (e.g., excessive disturbance caused by low-privacy office designs leads to irritation), health and well-being (e.g., strong glare can cause headaches) and work, emotional and physical stresses. Numerous studies have found that extreme temperature, lighting and noise can cause stress (Gmelch 1982; Leung et al. 2005; Selye 1956), especially for construction personnel who are required to work on-site under noisy, crowded and extremely hot or cold conditions for a set period of time. Although this does not affect one's job ability directly, poor site environment can affect construction personnel both emotionally (e.g., excessive noise causing irritation) and physically (e.g., poor site layout leads to unnecessarily long work path, resulting in the construction workers being physically driven out by walking to-and-fro repetitively). In addition, *poor home environment*, that is, poor functioning at home (e.g., insufficient free space and settings that ignore the importance of personal privacy), lowers performance on the job and, thus, induces work stress gradually (Organ 1979).

Lastly, construction sites are full of hazards, such as excessive noise from bulky equipment, working at a height, poor housekeeping and exposure to chemicals. Working in such an *unsafe site environment* induces work stress, as well as emotional and physical stress (Leung, Chan and Yu 2010). It also reduces construction personnel's attention to safety behaviours (Choudhry and Fang 2008; Sawacha, Naoum and Fong 1999) and has a direct impact on project performance.

4.3.1.3 *Interrelations between stressors*

Pearson correlation analysis was also used to test the interrelationships between the stressors (see Table 4.3). The results showed significant interrelationships between personal, interpersonal and task stressors, including the stressors of Type A behaviour (Ss1), work–home conflict (Ss2), poor interpersonal relationship (Ss3), work overload (Ss5), role ambiguity (Ss7), role conflict (Ss8), lack of job autonomy (Ss9) and effort–reward imbalance (Ss10). Distrust was only significantly related to poor interpersonal relationships, role conflict and lack of manpower support. Adequate safety equipment (Ss11) was not related to any personal stressors, but was significantly negatively correlated with the interpersonal stressor Ss3 and the task stressors Ss5, Ss7, Ss8, Ss9, Ss10 and Ss11.

1. *Type A behaviour.* Construction personnel with Type A behaviour tend to strive for achievement, behave competitively, act hostile and spend a lot time on their work (Glass 1977). Hence, it is readily understandable that their work would interfere with their family life. Due to their competitive, impatient and hostile characteristics, Type A construction personnel frequently argue with and overlook the feelings of people

who interact with them, which can easily lead to poor relationships (Yarnold and Grimm 1982). Construction projects are normally set up with limited periods of work for a particular quality goal. Thus, construction personnel suffer constant pressure to work, with few opportunities to relax. They have little time to carefully consider their role and responsibility, which usually comes with *role ambiguity* in construction project(s). It is difficult for them to keep harmonious *relationships* with other project members, especially in the face of *work overload* (Jamal 1990).

Furthermore, studies have found *effort–reward imbalance* will result in individual behaviour consistent with the Type A behaviour pattern, such as acting hostile or aggressive (Baron, Neuman and Geddes 1999). Construction sites often have *poor environments* in terms of noise, extreme weather and being crowded (Mansfield and Odeh, 1991). Prolonged exposure to such a difficult situation can easily lead to impatience, impulsivity and so on.

2. *Work–home conflict.* Construction personnel suffering from *work overload* and *poor interpersonal relationships among team members* usually need to pay a lot of time to their work and even work overtime. The higher their workload and the poorer their relationships, the more time and energy they need to put in. Therefore, construction personnel have less time to spend with their family and friends, which leads to work–home conflict. When facing *role ambiguity* and *lack of job autonomy*, construction personnel do not understand their roles very well in terms of responsibility, authority and performance expectations and they have less control over their jobs, which usually lowers performance. Hence, they need more time to finish their tasks. Similarly, *role conflict* for construction personnel could also lead to inefficient performance and is followed by more time put into their work, which, in turn, leads to work–home conflict. In the process of handling various stressors in construction projects, it is easy for construction personnel to feel three types of *stress (work, emotional and physiological)* when they are also from suffering work–home conflict.

    Construction personnel often complain about long working hours and unfair reward (i.e., *effort–reward imbalance*). Such imbalance between effort and reward gives construction personnel feelings of unfairness, reduces their motivation to work and finally aggravates work–home conflict. When *lacking feedback from superiors*, construction personnel cannot get prompt attention for their problems and they feel it is difficult to complete their work, which might prolong how long it takes to complete their work and induce work overload. This longer time spent working then induces work–home conflict. Moreover, construction personnel – especially construction workers – often work in a poor and unsafe site environment with extreme temperatures, bad air quality, noise and safety hazards. *Poor and unsafe site environments* induce physical and mental health problems, including headaches and stress (Mayo et al. 2012). The declining health of construction personnel further aggravates the conflict between their work and family life.

3. *Poor interpersonal relationships.* Construction personnel who have poor interrelationships with project team members may not want to communicate with other team members and their superiors, which hinders the establishment of trust among the team. Moreover, under such circumstances, their workload may be inappropriately allocated either via *overload* or *underload*. Once professionals suffer from work overload, they need to spend most of their time on their job and have little time to communicate with their team members, which may adversely affect their relationships with their colleagues. On the other hand, professionals with *role ambiguity* and *conflict* often question what they should be doing and whose requirements have high priority in the project team. They are concerned about whether they can meet all the demands or requirements from different parties, such as superiors, subordinates, colleagues, clients and regulations. It is easy for role ambiguity and conflict to lead to discontent of some team members and then poor interpersonal relationships follow. Moreover, construction personnel who have little control over their jobs (i.e., lack of job autonomy) usually encounter conflict with their supervisors over the job requirements and process, which has a negative impact on interpersonal relationships and the different types of *stress (work, emotional and physiological)*.

Construction personnel experiencing *effort–reward balance* usually have good moods and good job satisfaction, which, in turn, come with good interpersonal relationships (Lim and Ling, 2012). However, it is interesting that this study found a positive correlation between poor interpersonal relationships and *inadequate safety equipment*. Inadequate safety equipment normally induces stress for construction workers (Leung, Chan and Yu 2012; Leung, Chan and Yu 2011) and makes some feel anxious, which influences the communication among construction personnel and their interpersonal relationships.

In fact, it is difficult for construction personnel to maintain good communication with their superiors when they have poor interpersonal relationships with other team members, including colleagues and superiors. In that situation, superiors do not like to actively give professional feedback *(lack of feedback from superiors)*. On the other hand, lack of feedback not only causes construction personnel to be unaware of how their performance is and what they should do to improve, but it also may, to some extent, arouse discontent with the superior.

The results revealed that poor interpersonal relationships were positively correlated with *poor office environments, poor site environments and unsafe site environments*. Individuals with poor working relationship may have a higher chance of being assigned unfavorable job tasks in poor environment.

4. *Distrust.* The construction industry relies highly on manpower support. For example, architects need subordinates to draft detailed designs; surveyors need manpower for the preparation of Bills of Quantities; and project managers need clerks of works and supervisors to carry out the works on site. It maybe harder for individuals with distrust relationship with others to seek manpower support from the team or organization. Moreover, poor interpersonal relationships damage the communication and cooperation among construction personnel. Construction personnel without a clear understanding of their job responsibilities often feel *role conflict*. When conflict emerges, it is difficult for construction personnel to trust their colleagues, which thus induces *emotional stress*.

5. *Work overload.* Construction projects usually have lots of tasks and construction personnel often suffer from heavy workloads. In such a circumstance, construction employees must clearly understand their specific tasks or duties in each stage, such as investigating soil conditions and finalising building layouts before the structural calculations, checking client's requirements before making a detailed design, assessing contractual terms before signing the contract and checking the site space and construction programme before ordering materials. Otherwise, they may easily suffer from *role ambiguity* or *role conflict* because they do not know their responsibility exactly and some of tasks may contradict each other (e.g., cheaper materials or methods with low costs but poor workmanship). In some cases, they may need to rectify the tasks or the project by redoing some steps such as redesign, reordering, or reconstruction, which certainly induces work overload and increases their *stress (work, physiological and emotional)* in daily work. In fact, construction personnel with limited job autonomy are often required to handle many complicated tasks, which results in prolonged working hours. Since they have spent a lot of effort and time in the job, it is understandable for them to expect a high reward or otherwise perceive *effort–reward imbalance*. This can explain why the results showed that work overload was positively correlated with role ambiguity, role conflict, lack of job autonomy and effort–reward imbalance.

   Indeed, it is very difficult for construction personnel to perform complicated construction tasks well without any *feedback from their superiors* (either *praise, criticism* or *direction*). They may either waste lots of time on repetitive tasks to rectify problems or be nervous and wait for the feedback from superiors. This will directly affect work efficiency and induce work overload in a cumulative process. The provision of safety equipment is essential to avoid injuries for construction workers. Otherwise, an *unsafe environment* with *inadequate safety equipment* can increase the possibility of accidents and make the work process more complex, which induces work overload.

6. *Work underload.* On the other hand, construction personnel with the problem of *role ambiguity* or *conflict* may not know how to perform their tasks to satisfy contradictory demands at the same time. They may decide to simply follow a fixed procedure, do some simple tasks only, or even do nothing. Prolonged exposure to this type of situation will make construction personnel feel their job is boring and tedious, with no sense

of achievement that comes with finishing a challenging task. In this case, construction personnel can easily experience work underload, which can affect their *physiological stress,* such as loss of appetite, migraines and sleep disorders. Normally, construction companies' office environment is not good. Construction personnel who do not have enough work may find they are distracted by things other than work (e.g., pay attention on the quality of the office environment). This can explain why workload was positively correlated with poor office environment.

7. *Role ambiguity.* Since construction personnel suffering from *role ambiguity* cannot understand what *role* they are filling and what *authority* they have, it is inevitable that they will do some unrelated tasks in their daily work. Although they may put great effort in their work, their performance may not be satisfactory, resulting in limited *authority* being granted to them. For example, it is difficult or, arguably, impossible for an architect, who is employed by a contractor in a Design and Build Contract, to follow the instructions of the client (government or developer) and simultaneously consider the contractor's benefits comprehensively for all issues because there may be *conflict* between these two parties. Hence, such personnel may not consider their *reward to be fairly balanced with their effort,* which induces *stress (work, emotional and physiological)* in the implementation process.

Without any *feedback from superiors* or *career support from the organisation,* construction personnel hardly know exactly what kind of role they should play and how well they are performing. This means they cannot get enough information about their responsibilities, work content and scope from the superiors and future career direction or promotion in the company. Hence, they can easily suffer from role ambiguity.

An *unsafe site environment* easily causes accidents and incidents (Sawacha, Naoum and Fong 1999). In order to avoid dangers, construction personnel may need to take some additional actions, such as design and construction of temporary strutting, or they may need to pay additional attention to their jobs. Therefore, they may experience confusion and role ambiguity towards their original main tasks (e.g., excavation work and sloping work). Interestingly, *inadequate safety equipment* was found to be positively correlated with role ambiguity. Perhaps, construction personnel working with inadequate safety equipment are more likely to be distracted by the unsafe environment, which in turn leads to role ambiguity and a lack of clarity about job responsibilities.

The study also found a positive relationship between role ambiguity and organisational centralisation. Under a centralized organizational structure, Construction personnel at the management level often need to take control of their subordinates' tasks. In centralised organisations, it is quite common for construction personnel to manage multiple projects and tasks at the same time, which causes role ambiguity.

Moreover, crowded and noisy offices only provide limited space for the construction personnel. The office environment of small, developing firms are often poorer than that of large, developed firms, in which construction personnel working in such firms are most likely to play several roles and conduct several tasks (e.g., project manager, project coordinator and engineer), resulting in a higher possibility of role ambiguity. Meanwhile, some construction personnel have to work in a random rather than regular way on construction sites with poor physical environments. It maybe easier for them to suffer from role ambiguity due to the interchanging roles and tasks in both office and site.

8. *Role conflict.* In order to minimise the investment period, construction projects normally have to be completed within a limited period. Construction personnel may sometimes need to handle a number of projects at the same time, which can induce role conflict between competing urgent projects. Therefore, sufficient *manpower support* and *job autonomy* play important roles in their job. In fact, it is difficult for construction personnel with role conflict to satisfy all their 'roles' for the project that directly influence their *stress (work, emotional and physiological)*. Subsequently, this will affect their work performance – either the efficiency or the quality – and, in turn, the final reward from the project. Since the reward is determined by performance rather than effort, construction personnel with role conflict may find it easy to perceive their situation as having *imbalanced effort–reward*.

In most work settings, superiors determine the role of construction personnel. Superiors may assign several roles to construction personnel at a time and, sometimes, these roles are mutually conflicting (O'Driscoll and Beehr 1994). Moreover, a *lack of information from superiors* (e.g., information about the priority among different roles) can easily cause construction personnel to suffer from role conflict. The absence of *safety equipment* could expose construction personnel to risk of accidents and injuries. Undertaking risky tasks will cause injuries, but giving up these tasks will cause them to fail to fulfil their role, which can easily cause role conflict.

The results revealed that all four *environmental stressors (home, office, site and unsafe site)* are significantly related to role conflict, which fully corresponds to the traditional person–environment theory (Smithers and Walker 2000; see Chapter 3.1.1). Construction personnel cannot perform their roles perfectly and efficiently in *poor or unsafe working environments either in the office or on site* (e.g., extreme temperature and a chaotic environment). Moreover, a *poor home environment* may also affect their personal rest or sleep quality, which can influence their concentration on the job and induce role conflict while they are handling multiple tasks.

9. *Lack of job autonomy.* Construction personnel with limited job autonomy, *manpower support* and *feedback from supervisors* may take a long time to find potential solutions and complete each task by themselves. This can easily lead to perceptions of *effort–reward imbalance*. Job autonomy refers to people's freedom to manage their work, which can be related to the on-site working environment. *Poor and unsafe site environments* with *inadequate safety equipment* make it difficult for construction

personnel manage and control site works and for workers to implement construction projects.

10. *Effort–reward imbalance.* Provision of *safety equipment* not only protects construction personnel from injuries and accidents, but also demonstrates the concern of the organisation for its employees (MacDavitt, Chou and Stone 2007). In some sense, provision of safety equipment can be regarded as a benefit from the organisation to employees. If necessary safety equipment is absent, it is easy for construction personnel to perceive effort–reward imbalance. Similarly, recognition of supervisors is another kind of reward for construction personnel (De Jong et al. 2000). Absence of *feedback from supervisors* on the performance of construction personnel will cause disappointment among construction personnel and thus their reward will not be equivalent to their effort. Comfortable and even luxurious working environments have been treated as a means to attract and retain employees (Rowley and Jackson 2011). In contrast, *poor and unsafe site environments* upset construction personnel and communicate the indifference of management to the well-being of construction personnel. In this situation, construction personnel may think their effort is not valued or compensated.

11. *Inadequate safety equipment.* There is no question that it is important to establish a *tidy and safe environment* on a construction site with *adequate safety equipment* such as safety belts for workers working at height, high-visibility clothing for people working in high-traffic areas and safety helmets for everyone who enters the construction site (Dalby 1998). According to the Safety Manual issued by the Hong Kong government, all construction sites must employ a safety officer to manage safety issues and safety equipment on site and ensure a safe construction site (Development Bureau, 2000). In addition, there are safety regulations clarifying the duties of designers, clients and contractors in order to ensure occupational health and safety for the construction industry in the UK (Health and Safety Executive 2007). The designer has the responsibility to design for safety, while contractors and clients have to provide adequate safety equipment on site. It is important to get *feedback from experienced superiors*, especially for the compliances of working procedure or safety equipment provision with regulations.

12. *Lack of feedback from superiors.* This study found significant relationships between lack of feedback from superiors and *poor site environment* (e.g., hot and dusty) and *unsafe site environment* (e.g., sharp edges, nails on the ground and concrete blocks falling from the upper levels of the building). Feedback from experienced superiors is essential for construction personnel to design, manage and maintain a proper and safe site environment.

13. *Organisational centralisation.* Organisational centralisation relates to the organisational structure and determines how the organisation runs. Organisational centralisation affects the authority of construction personnel in their decision-making process. Hence, normally, higher

levels of organisational centralisation mean people have less control over the organisation. Under high levels of centralisation, construction personnel have fewer opportunities to join the decision-making process and obtain *career support* as they have less authority. Construction personnel are only expected to complete their tasks and they are restricted from other tasks in the organisation, which limits their career development.

14. *Lack of manpower support.* The construction industry emphasises teamwork. Thus, construction personnel without sufficient manpower support easily suffer from all of the *three types of stress: work, emotional and physiological.* Construction personnel suffering from lack of manpower support can either be junior members or members with poor interpersonal relationship with others. In both cases, individuals would have a higher chance of being assigned less favorable seats, such as having lesser privacy. Hence, they may have a higher change of suffering from *poor office environment.*

15. *Poor office and home environments.* Poor office and home environments include elements such as the size of people's accommodation and the noise and temperature (too hot or too cold) of the working and living environment (Ceylan, Dul and Aytac 2008). Senior construction personnel are often assigned a more favorable office space; meanwhile, due to their seniority, they may also be more capable in affording a more spacious home environment, vice versa. Hence, construction personnel are more likely to perceive the office environment as satisfying if they have a good home environment or vice versa. The study found that poor office environment was significantly related to all *three types of stress (work, emotional and physiological)*, while poor home environment was only related to work stress. This suggests that the office environment plays an essential role in the stress of construction personnel.

16. *Poor and unsafe site environment.* Poor site environments are noisy, crowded, too hot or too cold, too bright and so on (Petersen and Zwerling 1998). Construction personnel working in this sort of environment will have worse performance, including safety performance, than those working in a normal environment, since their work process can be significantly affected. For example, a brick mason working under very hot weather and bright sunlight will sweat heavily and be at risk of sunstroke (Morioka, Miyai and Miyashita 2006), which is certainly an unsafe environment and induces emotional and physiological stress. Moreover, construction personnel working in an unsafe site environment may have a larger gap between their actual and expected ability. Therefore, they experience more work stress.

4.4   Case Studies

4.4.1   Building Information Modeling in Housing Projects

This was a construction and development project undertaken in 2006 for a public housing estate in an international city. With a high degree of modernisation, the local built environment consisted of a small area with high population density and a number of high-rise residential buildings. Although there were other facilities on the estate, such as a playground, basketball court and shopping mall, the main focus of this project was the construction of eight 33-storey residential buildings, with similar and repetitive structures and designs.

4.4.1.1 Adoption of an Innovative Technology

The modern construction industry faces a technological change represented by the transition from CAD (Computer Aided Design)-based documentation to BIM. BIM takes architectural design to the next level by offering an expanded range of possibilities, due to the immense amount of information which can be encapsulated in and extracted from digital models. BIM will become the DNA of future construction projects. In addition to traditional 3D design modelling, BIM facilitates collaborative working among various project stakeholders in 4D (time), 5D (cost) and 6D (operations). A BIM-facilitated construction process allows the project team to make informed decisions in the early stages. Awareness of BIM and its use in construction is rapidly increasing in various countries, including the United States, Australia, the United Kingdom, Japan, Hong Kong and Singapore. Many construction professionals are beginning to gain familiarity with and use BIM. The trend towards its adoption in the sector is inevitable.

In view of the prominent advantages of BIM, the public sector developer of this project, whose vision included promoting BIM in the industry and implementing BIM in all its projects, decided to use this as a pilot for its introduction. The repetitive nature of the project was also one of the key reasons for its selection as a BIM pilot. Unlike AutoCAD, BIM is not just a visual aid but is in fact an information-processing system which demands input from all project participants from inception to completion. In other words, all the construction professionals involved, including architects, contractors and quantity surveyors, would be required to provide input to and be influenced by the output of BIM. However, most project participants did not acquire the necessary knowledge during their professional training, since AutoCAD remains the dominant computer-aided tool in use. Because of this, the developer for this project provided a series of BIM training events for the personnel involved to assist and encourage their adoption of the methodology.

However, the results of this training were unsatisfactory. Learning transfer was very low and the staff, instead of recognising the benefits of adopting BIM, experienced a high degree of stress in terms of learning this new technology, incorporating it into daily work, cooperating with each other by means of BIM and so on. Due to their different positions, they were affected by different sources of stress in regard to the learning and application of BIM (see Table 4.4).

Construction projects are often dynamic and demanding. In view of the requirement to acquire knowledge of and adopt the new BIM technology, the project in this case was particularly demanding for construction personnel. Construction personnel needed to spend extra effort and time on training and learning transfer. They were not only stressed by the excessive *workload*, but also the *home–work* conflict induced by the excessive amounts of work and the unfamiliar technology. For instance, the project manager indicated

Table 4.4 Stressors experienced by construction personnel involved in the BIM project.

Stressors	Construction personnel
Personal	
Type A personality	PM: I have a strict requirement for efficiency and expect my subordinates to complete their jobs as quickly as they can. However, since BIM is a new technology, the progress of my subordinates in this project was really slow and this also affected the progress of our other project. My subordinates were all afraid of me since I easily **lost my temper** due to the time pressures.
	Eng: This is not the first time I have been involved in a public housing project. With this previous experience, I should have completed the building services design **in a shorter time**. However, since I am not familiar with BIM, I encountered a lot of problems transforming the AutoCAD 2D drawings into the BIM format. It was really **time consuming**. Every time I was stuck and encountered difficulties with the software, the question 'Is it really necessary to use BIM when it is so time consuming?' crossed my mind.
Work–home conflict	Arch: Although I am glad that the BIM training was provided, the series of courses certainly slowed down my work progress, which resulted in frequent overtime during that period . . . **my wife complained** to me about this.
	Eng: Since I had to spend most of my time transforming the 2D drawings into 3D BIM models, I **worked until midnight** the whole week before submission deadlines.
Interpersonal stressors	
Poor interpersonal relationships	PM: I always feel that I am alone in the project team although I am the project manager . . . I was regularly **questioned by my subordinates** about the necessity of adopting BIM. Although I myself also had doubts, I have to meet the client's expectations and so I asked my subordinates to work on BIM. I feel that project team members **seem to dislike me**.
Distrust	CW: When we need to redo some construction work, I always think it is because of **other people's mistakes** in design and planning, due to their unfamiliarity with BIM
Task	
Work overload	PM: I am often stressed by the complicated coordination and difficulties involved in balancing tasks and demands between the client and the project team . . . For instance, the client provided BIM training for all project team members. However, it was not valued by some of them. Some of the members considered BIM time consuming and unnecessary and thus refused to learn it. However, as a project manager, I am responsible for ensuring that the project runs smoothly and achieves the client's expectations. I **had to spend a lot of time** convincing other team members to learn and apply BIM. In addition, I often need to handle **more than one project** at the same time. This BIM project has slowed down my whole schedule. I don't have time for a rest.
	Eng: Although BIM can help with identifying clashes in the early design stage, a comprehensive set of information input by every related party is necessary. Hence, we suffered from **heavy workload** in the early stages of the project, which is different from a traditional project.
Role conflict	PM: I need to **complete the project as required by the client in terms of cost and time**, while my project team members **ask to delay** project milestone completion dates because of this new technology.

Table 4.4   *(Continued)*

Stressors	Construction personnel
	QS: Although I find that **BIM is really powerful** and speeds up the estimation process significantly (disregarding the configuration stage), I personally was a bit **hesitant about its popularity**. As you may know, the work nature of QS is a bit mechanistic. The adoption and popularity of BIM may one day kill professionalism in QS.
Role ambiguity	Eng: We are supposed to take responsibility for the building services design only. However, there is a series of information sharing and updating processes required in the BIM project (consolidation of a BIM database). **I really don't know my role and responsibilities** in such a complicated system.
Lack of feedback from superiors	Eng: I reported to my superior about my BIM learning outcomes, the application of BIM in the design process and the results … however, he **did not say** whether the results were good or bad or make any suggestions for improvements. Perhaps he was not that familiar with BIM either.
Effort–reward imbalance	Arch: I have to spend lots of my spare time learning BIM, so that I can use it to finish my job within the required time. However, I **get only a small symbolic award** … It is very unfair.
Inadequate safety equipment	Eng: Although safety equipment like a safety helmet is provided, it may **not be in good condition**. Items are sometimes **old and dirty**. I doubted their functionality. I tried to bring my own safety helmet instead.
Organisational stressors	
Organisational centralisation	PM: I am always responsible for **decision making** about **all** the important issues in the construction project. The compulsory application of BIM increases the difficulty of my job.
	QS: Since BIM is very new to me and my colleagues, my firm has been paying extra attention to our work in this trial project. I am **closely monitored** by my immediate leader, which makes me a bit nervous.
Organisational formalisation	PM: Our company has **strict standards and procedures** for the documentation of drawings, especially those for submission to the government. The government requires 2D submissions with specific standards. We have to convert any 3D BIM model developed into 2D drawings for submission, which is a very tedious process.
Lack of staffing support	Arch: I usually find someone to help with the typical design work, especially when deadlines are approaching. However, since my subordinates were not experienced in BIM either, their **support** was really **limited**. I can only contact the training firm when I encounter any BIM problems at work … However, it was also not easy for me to explain the difficulties I encountered to the BIM trainers.
	Eng: I really need many workers to install the equipment. However, such trade **workers** always seem to be **in short supply**.
Physical stressors	
Poor office environment	Arch: My workplace is very **messy and hot**.
	Eng: I cannot stand the **low temperature** in my office – it is **too cold**.
Poor on-site environment	PM: When I inspect the site, I have to endure the **harsh noise** made by the vibrator and piling equipment.
	Arch: I usually find that my construction site is full of **wastage**.
	CW: As a concretor, I usually work outside under the **strong sunshine** in summer or **fierce winds** in winter … I witnessed a worker fall down due to heat stroke last year … I am still afraid sometimes when I think of this.
Poor home environment	Arch: There is a highway outside my flat which is too **noisy**. I really cannot sleep at night.
	Eng: My house is too **small** and I live with my parents.

PM = Project manager; Arch = Architect; Eng = Engineer; QS = Quantity surveyor; CW = Construction worker.

that 'the client provided BIM training for all project team members. . . . Some of them considered BIM time consuming and unnecessary and thus refused to learn it. I had to spend a lot of time convincing other team members to learn and apply BIM'. On the other hand, the architect also reported that 'although I am glad that BIM training was provided, the series of courses certainly slowed down my work progress, which resulted in frequent over-time during that period. My wife complained to me about this'. Since BIM was new to the project team, their efficiency dropped significantly, which was especially problematic and stressful for those with Type A personalities, who were often driven by their impulse to act and react in the fastest possible manner. In addition, due to the uncertainties involved in using the new tech-nology, the project team members failed to align with the BIM adoption, resulting in poor interpersonal relationships and distrust within the team. As the project manager put it, 'I was always questioned by my subordinates about the necessity of adopting BIM. Although I myself also had doubts, I have to meet the client's expectations and so I asked my subordinates to work on BIM. I feel that project team members seem to dislike me'.

BIM is a new information-processing system which brings construction project management into a new era. Its adoption not only results in the visu-alisation of drawings, but also changes the roles and practices of construction personnel. As the engineer explained, 'we were supposed to take responsibility for the building services design only. However, there is a series of information sharing and updating processes required in the BIM project (consolidation of a BIM database). I really don't know my role and responsibilities in such a complicated system'. Since BIM was still new to the project team and the construction sector as a whole, it was difficult for them to get *manpower support* from their firm, or even to obtain concrete feedback or support from their superiors. 'I reported to my superior about my BIM learning outcomes, the application of BIM in the design process and the results . . . however, he did not say whether the results were good or bad or make any suggestions for improvements. Perhaps he was not that familiar with BIM either', suggested the engineer. On the other hand, they had to invest extra effort in this project in comparison with other work, without a corresponding increase in reward. As the architect explained, 'I have to spend lots of my spare time learning BIM, so that I can use it to finish my job within the required time. However, I get only a small symbolic award . . . It is very unfair'. Such an *effort–reward imbalance* can result in negative emotions and even an increased intention to leave the organisation (De Jong et al. 2000).

In the process of innovation adoption, organisational structures, which determine the degree of autonomy and the flexibility of conducting tasks in different ways, are key (Andrews, George and Richard 2009). However, some of the construction personnel involved in this project indicated that their organisations were *centralised* and *formalised*. For instance, the quantity sur-veyor explained that 'since BIM is very new to me and my colleagues, my firm has been paying extra attention to our work in this trial project. I am closely monitored. . . .' On the other hand, the project manager also stated that his company had 'strict standards and procedures for the documenta-tion of drawings, especially those for submission to the government. The government requires 2D submissions with specific standards. We have to convert any 3D BIM model developed into 2D drawings for submission, which is a very tedious process'. An inappropriate organisational structure not only causes stress for construction personnel, but also hinders innovation in the sector.

Lastly, as indicated by previous empirical studies (see e.g. Driskell and Salas 1996; Goldenhar, Williams and Swanson 2003; Leung et al. 2008), these construction personnel also suffered from stress induced by the *poor office, site and* even *home* environments. Early studies on the person–environment fit demonstrate there is a close association between people and their surroundings. For construction personnel, particularly those working on site, the unsafe outdoor site environment may be one of their most significant stressors. For instance, the construction worker in this project commented that 'as a concretor, I usually work outside under the strong sunshine in summer or fierce winds in winter . . . I witnessed a worker fall down due to heat stroke last year . . . I am still afraid sometimes when I think of this'. On the other hand, the *provision of safety equipment* is also a concern. Although this is governed by regulations, the quality of such equipment can be questionable. As the engineer explained, 'although safety equipment like a safety helmet is provided, it may not be in good condition. Items are sometimes old and dirty. I doubted their functionality. I tried to bring my own safety helmet instead'.

4.4.2   Design of a Water Tank in a Residential Project

A water storage tank is an important part of a water distribution system in any residential project. Water can be pumped into the tank in non-peak periods and pumped out to serve residents during peak periods. In this 30-storey residential project, the designer initially designed a ground tank and located it in the basement of the building at a corner of the underground car park. In accordance with the design, the construction of the car park and the installation of the water tank were completed, since the basement was scheduled in the early construction stages, immediately after completion of the foundation. However, the client then complained that there were insufficient parking spaces available for future residents. As they intended to sell these spaces, this would have reduced their profits. The designer was therefore instructed to move the tank.

An elevated tank (i.e., located on the roof) was subsequently designed with a view to allowing the basement space to be used more profitably (such as for car parking) and lower the long-term cost of continuously pumping water up through the distribution system (since water can be distributed by the pressure maintained by gravity in an elevated tank). This plan was sent to the contractor to implement the relocation of the water tank. The project was already in its latter stages, so this caused a delay to the schedule. In order to speed up the process, the contractor removed the tank that had already been installed in the basement, while at the same time feeding back to the designer and client that relocation to the roof would incur a significant

increase in the costs of the initial work (lifting and installation) and maintenance (since ground tanks facilitate water quality testing in the operational
stages and are also safer). The change was also said to decrease the aesthetic
values of the project since it is visually preferable for the water tank to be
hidden in the basement. The completion of the building, especially the top
floors, slowed down while a decision on the water tank was made.

The clients eventually decided to revert to the first design; that is, the
basement water tank, so as to avoid the high installation and maintenance
costs and to prevent further delays. During the process, the designers, quantity surveyors and project managers all faced different types of stressors
(see Table 4.5) arising from the difficulties in communication with the client,
the repetitive works quantity, the reschedule of the construction programme
in a limited timeframe and so on.

A construction project is a complex process, not a simple linear formula of
$aX + bY = cZ$. It is full of uncertainties and change. In this case, the project
team was facing the dilemma of whether to move the water tank to the roof
or leave it in the basement. Whenever there were changes, the project progress
would be suspended due to the need to rework the design, estimation,
programming, construction and so on. This resulted in *extra workload* for
the project team members. As the project manager stated, 'the repetitive construction, removal and replacement work with the water tank really drove me
crazy . . . This is just one of the many projects that I have to handle. Whenever
they announced a change, I had to start all over again with the construction
programme, human resources allocation, method statements and so on . . .
My workload was increased unreasonably'. Such work overload means overtime for construction personnel, resulting in *work–home conflict*. As one of
the engineers explained, 'every time I was instructed to change the design, I
had to spend a long time getting a whole new set of reasonable design drawings done, even though I was at home'. Moreover, the delay in progress would
often cause excessive stress, especially for personnel with *Type A personalities*,
who are often keen to complete their tasks as soon as possible.

In this case, the repeated changes of the water tank design and location
caused dissatisfaction for the project team members. The project manager
disagreed with the changes, but his opinions were not accepted. 'To be honest,
I disagreed with the stupid idea of moving the water tank from the basement
to the roof. I raised this at the very beginning. However, the designer simply
ignored it . . . That's why I contacted the client directly . . . I know this pissed
the designer off', he explained. Although the project manager took this view,
he nevertheless started out by following instructions. The *intra-role conflict*
he suffered caused him frustration. On the other hand, the architect thought
that the project manager had stepped across his boundary and called into
question his professionalism. The change had induced conflict between
members of the project team, causing *poor interpersonal relationships* and
distrust and resulting in poor cooperation in the latter stages.

Furthermore, since the quantity surveyor in this project was a junior
professional, she was assigned only simple estimating tasks, resulting in
boredom. She explained that 'As a junior QS, my main role is to measure the
quantities, which is simply repetitive and tedious . . . I think I will be able to

Table 4.5 Stressors experienced by construction personnel involved in the water tank project.

Stressors	Construction personnel
Personal	
Type A personality	PM: I have been trying my very best for a long time to ensure project completion on time … The project has already been delayed for two months. However, the client started to struggle with the decision of whether to go for the original (basement water tank) or the new plan (elevated water tank) at that critical point. I could only wait for their decision while at the same time **doing as much as** I could to make sure that the project was completed on time. QS: I sometimes **worried about** my valuations even when I am at home.
Work–home conflict	PM: I was really frustrated to learn that the clients wanted to change the design again. Due to the tight project timeframe, I was forced to **work at night** revising the construction programme and planning the dismantling and installation sequences. Eng: It was not like playing with Lego, where we could simply move the tank from the ground to the roof … Every time I was instructed to change the design, I had to spend a **long time** getting a whole new set of reasonable design drawings done, even though I was at **home.**
Interpersonal stressors	
Poor interpersonal relationships	PM: To be honest, I disagreed with the stupid idea of moving the water tank from the basement to the roof. I raised this at the very beginning. However, the designer simply ignored it. That's why I contacted the client directly … I know this **pissed the designer off.** QS: As a junior quantity surveyor, I am not familiar with the difference between a ground and elevated water tank … However, I found that I had nobody to turn to for help when having a problem in estimation … **Nobody wanted to help** me, because they thought that this was just a small and easy task.
Distrust	Arch: I don't think that the contractor has any expertise in aesthetic issues … Their **opinions** on designs were actually **not necessary.** CW: We don't understand why so much redundant work was required in this project. There must have been some mistakes made by incompetent professionals in the project team … **They were just trying to cover** it up.
Task	
Work overload	PM: The repetitive construction, removal and replacement work with the water tank really drove me crazy … This is just one of many projects that I have to handle. Whenever they announced a change, I had to **start all over again** with the construction programme, human resources allocation, method statements and so on … My **workload was increased** unreasonably. Eng: The client and architect often struggled to balance the aesthetic and profitability aspects of the water tank plan. They kept changing their minds all the time. It was really **demanding for us** to amend the design within such a short period.
Work underload	QS: As a junior QS, my main role is to measure the quantities, which is **simply repetitive and tedious** … I think I will be able to contribute much more than these boring tasks later on when I have amassed more practical experience. The procurement process is more challenging for me. CW: I am an ironworker … my job is to fasten the iron steel using the same action all day long … It is so **boring!**

(Continued)

Table 4.5 (Continued)

Stressors	Construction personnel
Role conflict	PM: As a project manager, I have a heavy responsibility for ensuring project completion on time. However, the client and designer kept changing the project design, which slowed down the project... Although I **did not agree** with their ideas, I **still had to follow** their instructions... Seriously, it was quite frustrating.
	CW: As a water worker, my professional role is to install the water tank in a standard manner, However, the foreman usually asks me to **do some other things** during slack work times, like buying cigarettes and other things.
Lack of job autonomy	PM: As a project manager, all my job responsibilities are constrained very clearly in the contract... I **cannot manage the project my way**, even though it may be more efficient.
	Arch: I feel that I am like a robot... I can only do my design job in accordance with the instructions of the client... and **cannot express my own ideas.**
Lack of feedback from superiors	QS: As a junior quantity surveyor, I don't have much estimation experience... and would like to have had comments on my performance from my boss... But she **did not give any feedback**... I worried about that and did not know what was going on.
Organisational stressors	
Organisational formalisation	Eng: It is the **policy** of our organisation to report and record all amendments of works, no matter how minor... It is especially time consuming for me to do this documentation and report works for projects with many variations, like this one.
Lack of career support	PM: The company **does not provide any training courses** about project management knowledge and skills.
	Eng: My company **does not provide the engineers with professional development consulting assistance.**
	QS: I do **not clearly understand my career development in the industry.** Since I don't have much contact with my seniors, I am not sure about the career path in this organisation.
Lack of financial support	Arch: Although I was asked to work overtime to complete the architectural design work, I **got no extra compensation.**
	QS: My company would **not pay the fee for a seminar**, even though the event was related to the latest estimation software.
Physical stressors	
Poor office environment	PM: It is common for contractors to rent apartments in factory buildings as offices... Since they are not designed for use as office buildings, **natural light is rare and ventilation is not good**... **Hygiene is also a big problem.**
	QS: Although the government has been promoting the constant indoor temperature of 24.5C for a long time, the **temperature of our office is always around 20.** It is ridiculous... I have to wear a coat whenever I am in.
Poor site environment	PM: The site is always a **mess.** Even though we order the foreman to tidy it up from time to time, especially before important site visits, it will go back to being a mess after a few days.
	Eng: I have to conduct site visits for regular checks. Every time I went home after a site visit, there was a **thick layer of dust** on my face... You can imagine how dusty the site is.
	CW: In entering this industry, we should be able to endure the **extreme outdoor** environment. However, it's really very **hot and humid** during the summer.
Unsafe site environment	Eng: There are many **iron nails scattered around on the site**, which could easily **hurt** people.
	CW: Construction **waste is always dropped from height**... My colleague's friend was killed by a hammer falling from height two years ago. My wife worries about me.

PM = Project manager; Arch = Architect; Eng = Engineer; QS = Quantity surveyor; CW = Construction worker.

contribute much more than these boring tasks later on when I have amassed more practical experience. The procurement process is more challenging for me'. *Work underload* can also cause stress since it does not allow the individual to demonstrate his or her ability. *Feedback from superiors* is also especially important for young professionals (CIOB 2006). However, the junior QS said that she had not been given such feedback, so she felt unclear about her work direction. In contrast, however, senior professionals require *job autonomy*. As the architect put it, 'I feel that I am like a robot . . . I can only do my design job in accordance with the instructions of the client. . . and cannot express my own ideas'. The lack of autonomy resulted in frustration with his career progression.

Organisational structure has a direct impact on the work practices of individual employees. In a *formalised organisational structure*, individuals are forced to carry out various bureaucratic tasks. As the engineer explained, 'it is the policy of our organisation to report and record all amendments of works, no matter how minor . . . It is especially time consuming for me to do this documentation and report works for projects with many variations, like this one'. A rigid organisational structure can hinder the progress of a construction project. As well as structure, support from the organisation, in terms of both *career support* (such as training for professional development) *and financial support* (such as subsidy for professional training events) are essential ways for the firm to express its values towards its employees. This may be especially important in retaining staff who are under excessive stress.

Lastly, the construction personnel in this project also described suffering from physical stressors arising from the *poor office and site environment*. This is, again, especially true for construction workers, who are forced to work in a poor and unsafe environment. As one of them put it, 'In entering this industry, we should be able to endure the extreme outdoor environment. However, it's really very hot and humid during the summer. Construction waste is always dropped from height . . . My colleague's friend was killed by a hammer falling from height two years ago. My wife worries about me'. On the other hand, although construction professionals, as distinct from site workers, are less likely to have to spend time on site and deal with extreme weather conditions, they are still affected by poor office environments in terms of temperature fluctuations, poor ventilation and insufficient lighting.

4.5   Practical Implications

Based on the stressor comparison results, several practical implications can be provided. Young construction personnel are strongly recommended to communicate with their experienced seniors about their needs so as to lower their chances of suffering from work underload and lack of job autonomy. Construction companies need to draw up clear performance evaluation methods for their staff, especially young staff, in order to avoid an imbalance between effort and reward.

To manage construction personnel's stress, their personality (e.g., Type A behaviour) and personal life should be carefully handled. Although people's personality and personal life cannot be manipulated, it is suggested that management personnel identify and understand the different personalities and personal lifestyles of employees via personality tests and family background surveys when they first enter the organisation. In this way, they can provide tailored training or support. Senior managers are also advised to conduct regular stress measurement of construction personnel so that they can understand the nature and degree of their work stress. Thus construction companies have to discourage the practice of overtime among their staff so as to maintain a balance between employees' work and family life (Leung et al. 2008). Family members of construction personnel are encouraged to participate in informal social gatherings outside office hours, such as Christmas parties and sports in order to facilitate an overall supportive social environment (Leung, Chan and Yu 2009).

To improve interpersonal relationships, team activities are strongly recommended, such as team-building camps, site visits and study tours. By participating in these team activities, construction personnel have chances to know more about other colleagues and strengthen their relationships. It is also suggested that training opportunities on communication skills and team-building techniques be provided to construction personnel in order to help them cooperate and trust each other more. Furthermore, since distrust among project team members/colleagues may lead to poor communication, it is suggested that young personnel talk with their superiors if they face problems that they cannot deal with and proactively invite their superiors to evaluate their work and suggest how to improve their performance.

Since *work overload* and *underload* are both essential problems for the construction personnel, regular progress meetings are highly recommended. During progress meetings, construction personnel should not only report their work progress, but also express their workload difficulties (both overload and underload). In this way, employers can clarify their specific roles and job allocation of individual professionals. In addition, as work load is one kind of foreseeable risk to physical and mental health, employers have a duty of care in protecting construction personnel from risks of harm, including that of psychiatric damage (see *Walker vs. Northumberland County Council* (1995) for a successfully sued case with abovementioned context; Cooke 2009).

Because construction personnel are constantly handling multiple tasks (including both complicated and simple tasks at the same time), it is suggested that construction personnel suffering from work overload develop team spirit and trust in other team members so that they can delegate part of their tasks to their team members or subordinates if needed (Gryna 2004). In this way, the subordinates can help finish simple work and these construction personnel can devote their full attention to the most important and complex work. Therefore, they will be much less likely to suffer from work overload. For work underload, construction companies need to communicate with personnel frequently and share visions and expectations about workload (Leung et al. 2008).

This study identified role conflict, role ambiguity and job autonomy as major task stressors. It is thus suggested that organisations specify the responsibilities of the construction personnel. Moreover, it is suggested that superiors place trust in the performance of construction personnel. Superiors should not require them to do everything strictly under their supervision of the work process. In addition, it is suggested that construction personnel suffering from a lack of job autonomy assess the work environment for ergonomic problems that may need correction, address problems with commuting to and from work and seek support from family members when needed (Lacaille et al. 2004).

Organisational centralisation influences the way construction personnel complete tasks. It is strongly recommended that organisations develop clear organisational structure and specify the responsibilities for each position (Ferris et al. 1996). Moreover, construction personnel should be empowered to the operational problems in the construction process. It is recommended that documentation and paperwork be reduced to simplify complicated procedures.

To resolve stressors in the organisational support domain, it is important that construction companies provide sufficient support from supervisors and manpower. It is recommended that companies provide appropriate safety equipment to reduce the possibility of accidents and injuries. To balance effort and reward, it is suggested that organisations establish performance-evaluation methods and a path for career development of construction personnel. Since many construction personnel feel it is unfair if they get little reward for working overtime, it is strongly suggested that organisations specify the wage per hour for overtime work and follow it strictly. It is also suggested that organisations negotiates with employees for a reward system satisfying by both parties (Kuper et al. 2002).

Working in a *poor office and site environment* will also induce a lot of stress, especially for young construction personnel. It is suggested that organisations provide construction personnel a comfortable environment, either in the office or on site. Although air-conditioners are often installed in offices, the temperature in offices is often too low nowadays. Hence, it is suggested that office temperatures be maintained at a constant and moderate level, such as 23–27°C, as suggested by the HKSAR Government (Architectural Services Department 2012). On the other hand, lighting systems (both natural and artificial) should also be managed well so as to prevent insufficient lighting and glare (Leung, Chan and Yu 2009). To avoid creating overcrowded environments, it is expected that organisations will keep a proper width between desks to ensure the privacy of construction personnel (Oldham and Rotchford 1983). Since too much noise will decrease employees' work performance, it is recommended that construction stakeholders put posters around the office, reminding and encouraging staff not to speak loudly when communicating with others face to face or over the telephone. In addition, it is also recommended that the office environment be kept clean. It is recommended that the office cleaner clean the office regularly and gather the rubbish and waste paper frequently. For the on-site environment, it is suggested that organisations stagger work hours (e.g., work in the early morning and night, with a break during noon time) in hot weather to decrease the risk of heatstroke (Morioka, Miyai and Miyashita 2006). Construction stakeholders are recommended to enforce the rules on personal protective equipment (PPE) strictly, so as to protect workers/professionals on site from dangers such as falling objects, dust and excessive noise. During dry weather, it is also suggested that water be sprayed on site to decrease dust. Wall and ceiling panels with acoustic insulation are essential to minimise noise for construction personnel who work on site (Leung et al. 2005). Some construction firms would construct temporary landscape and green roof for on-site office for dust and heat insulation; in which these greenery will be part of the final building product, creating no waste.

References

AbuAlRub, R. (2004) Job stress, job performance and social support among hospital nurses. *Journal of Nursing Scholarship*, 36, 73–78.

Ahuja, M.K. and Thatcher, J.B. (2005) Moving beyond intentions and toward the theory of trying: Effects of work environment and gender on post-adoption information technology use. *MIS Quarterly*, 29(3), 427–459.

Andrews, R., George, A.B. and Richard, M.W. (2006) Subjective and objective measures of organizational performance: An empirical exploration. In A.B. George, J.M. Kenneth, J. Laurence and R.M.W. O'Toole (eds), *Public Service Performance: Perspectives on Measurement and Management*. Cambridge: Cambridge University Press.

Architectural Services Department (ASD) (2012) *General Specification for Air-Conditioning, Refrigeration, Ventilation and Central Monitoring and Control System Installation in Government Buildings of the Hong Kong Special Administrative Region*. Retrieved from https://www.archsd.gov.hk/media/11446/e217.pdf

Baron, R.A., Neuman, J.H. and Geddes, D. (1999) Social and personal determinants of workplace aggression: Evidence for the impact of perceived injustice and the Type A Behavior Pattern. *Aggressive Behavior*, 25, 281–296.

Bass, B.M. (1999) Two decades of research and development in transformational leadership. *European Journal of Work and Organizational Psychology*, 8(1), 9–32.

Beehr, T.A. (1995) *Psychological Stress in the Workplace*. London:, Routledge.

Beehr, T.A. and Bhagat, R.S. (1985) *Human Stress and Cognition in Organizations: An Integrated Perspective*. New York: Wiley.

Beehr, T.A. and Jex, S.M. (2001) The management of occupational stress. In C.M. Johnson, W.K. Redmon and T.C. Mawhinney (eds.), *Handbook of Organizational Performance*, 51–80. New York: Haworth Press.

Bhanugopan, R. and Fish, A. (2006) An empirical investigation of job burnout among expatriates. *Personnel Review*, 35(4), 449 – 468.

Bowen, P., Edwards, P., Lingard, H. and Cattell, K. (2014) Predictive modelling of workplace stress among construction professionals. *Journal of Construction Engineering and Management*, 140(3).

Bresnen, M.J., Bryman, A.E., Ford, J.R., Beardsworth, A.D. and Keil, E.T. (1986) The leader orientation of construction site managers. *Journal of Construction Engineering and Management*, 112, 370–386.

Buck, V. (1972) *Work Under Pressure.*, London: Staples Press.Caplan, R.D. and Jones, K.W. (1975). Effects of work load, role ambiguity and Type-A personality on anxiety, depression and heart rate. *Journal of Applied Psychology*, 60, 713–719.

Carayon, P. (1992) A longitudinal study of job design and worker strain: preliminary results. In J.C. Quick, L.R. Murphy and J.J. Hurrell (eds), *Stress and Well-being at*

Work: Assessment and Interventions for Occupational Mental Health, 19–32. Washington, DC: American Psychological Association.

Ceylan, C., Dul, J. and Aytac, S. (2008). Can the office environment stimulate a manager's creativity? *Human Factors and Ergonomics in Manufacturing*, 18(6), 589–602.

Chang, E.C., Maydeu-Olivares, A. and D'Zurilla, T.J. (1995) Optimism and pessimism as partially independent constructs: Relationship to positive and negative affectivity and psychological well-being. *Personality and Individual Differences*, 23(3), 433–440.

Chang, T.C. and Ibbs, C. (1990) Priority ranking – A fuzzy expert system for priority decision making in building construction resource scheduling. *Building and Environment*, 25(3), 253–267.

Chinowsky, P.S., Diekmann, J. and O'Brien, J. (2010) Project organizations as social networks. *Journal of Construction Engineering and Management*, 136(4), 452–458.

Chiocchio, F., Forgues, D., Paradis, D. and Iordanova, I. (2011) Teamwork in integrated design projects: Understanding the effects of trust, conflict and collaboration on performance. *Project Management Journal*, 42(6), 78–91.

Child, J., Faulkner, D. and Tallman, S.B. (2005) *Cooperative Strategy: Managing Alliances, Networks and Joint Ventures*. New York: Oxford University Press.

Chan I.Y.S., Leung M.Y. and Yu, S.W. (2012) Managing stress of Hong Kong expatriate construction professionals in Mainland China: A focus group study to exploring individual coping strategies and organizational support. *Journal of Construction Engineering and Management*, 138(10), 1150–1160.

Chartered Institute of Building (CIOB) (2006) *Occupational Stress in the Construction Industry*. Berkshire: CIOB.

Choudhry, R.M. and Fang, D. (2008) Why operatives engage in unsafe work behaviour: Investigating factors on construction sites. *Safety Science*, 46(4), 566–584.

Cooke, R. (2009) *Planning, Measurement and Control for Building*. Chichester: John Wiley.

Cooper, C.L. (2001) *Organization Stress: A Review and Critique of Theory, Research and Application*. Thousand Oaks, CA: Sage.

Cordes, C.L. and Dougherty, T.W. (1993) A review and an integration of research on job burnout. *Academy of Management Review*, 18(4), 621–656.

Cox, T. and Griffiths, A.J. (1995) The assessment of psychosocial hazards at work. In M.J. Shabracq, J.A.M. Winnubst and C.L. Cooper (eds.), *Handbook of Work and Health Psychology*. Chichester: Wiley.

Cropanzano, R., Howes, J.C., Grandey, A.A. and Toth, P. (1997) The relationship of organizational politics and support to work behaviors, attitudes, and stress. *Journal of Organizational Behavior*, 22, 159–180.

Dalby, J. (1998) *EU Law for the Construction Industry*. Oxford: Blackwell Publishing.

Davidson, M.J. and Sutherland, V.J. (1992) Stress and construction site managers: Issues for Europe 1992. *Employee Relations*, 14(2), 25–38.

De Jong, J., Bosma, H., Peter, R. and Siegrist, J. (2000) Job strain, effort–reward imbalance and employee well-being: A large-scale cross-sectional study. *Social Science and Medicine*, 50(9), 1317–1327.

Dench, S., Aston, J., Evans, C., Meager, N., Williams, M. and Willison, R. (2002) *Key Indicators of Women's Position in Britain (Women & Equality Unit)*. London: Department of Trade and Industry.

Development Bureau (2000) *Construction Site Safety Manual*. Retrieved from http://www.devb.gov.hk/en/publications_and_press_releases/publications/construction_site_safety_manual/index.html

Djebarni, R. (1996) The impact of stress in site management effectiveness. *Construction Management and Economics*, 14(4), 281–293.

Djebarni, R. and Lansley, R. (1995) Impact of site managers' leadership on project effectiveness. *Proceedings of 1st International Conference on Construction Project Management*, Singapore, 123–131.

Driskell, J.E. and Salas, E. (1991) Group decision making under stress. *Journal of Applied Psychology*, 76, 473–478.

Driskell, J.E. and Salas, E. (1996) *Stress and Human Performance*. Hillsdale, NJ: Erlbaum.

Eagle, B., Miles, E. and Icenogle, M. (1997) Interrole conflicts and the permeability of work and family domains: Are there gender differences? *Journal of Vocational Behavior*, 50, 168–184.

Eisenberger, R., Huntington, R., Hutchison, S. and Sowa, D. (1986) Perceived organizational support. *Journal of Applied Psychology*, 71, 500–507.

Emslie, C., Hunt, K. and Macintyre, S. (2004) Gender, work–home conflict and morbidity amongst white-collar bank employees in the UK. *International Journal of Behavioral Medicine*, 11(3), 127–134.

Eriksson, C.B., Bjorck, J.P., Larson, L.C., Walling, S.M., Trice, G.A., Fawcett, J., Abernethy, A.D. and Foy, D.W. (2009) Social support, organizational support, and religious support in relation to burnout in expatriate humanitarian aid workers. *Mental Health, Religion and Culture*, 12(7), 671–686.

Ferris, G.R., Frink, D.D., Galang, M.C., Zhou, J., Kacmar, K.M. and Howard, J.L. (1996) Perceptions of organizational politics: Prediction, stress-related implications and outcomes. *Human Relations*, 49(2), 233–266.

Fielder, F.E. and Garcia, J.E. (1987) *New Approaches to Effective Leadership: Cognitive Resources and Organizational Performance*. Chichester: John Wiley.

Fisher, C.D. and Gitelson, R. (1983) A meta-analysis of the correlates of role conflict and ambiguity. *Journal of Applied Psychology*, 68, 320–333.

Francesco, A. and Gold, B. (2005) *International Organizational Behavior*. New Jersey: Pearson Prentice Hall.

French, J. and Caplan, R. (1970) Psychosocial factors in coronary heart disease. *Industrial Medicine and Surgery*, 39, 383–397.

French, J. and Caplan, R.D. (1972) Organizational stress and individual strain. In A.J. Marrow, (ed.), *The Failure of Success*. New York: AMACOM.

Friedman, M. and Rosenman, R.H. (1974) *Type A Behaviour and Your Heart*. New York: Alfred Knopf.

Frone, M.R., Russell, M. and Cooper, M.L. (1992) Antecedents and outcomes of work–family conflict: Testing a model of the work–family interface. *Journal of Applied Psychology*, 77, 65–78.

Ganster, D.C., Fusilier, M.R. and Mayes, B.T. (1986) Role of social support in the experience of stress at work. *Journal of Applied Psychology*, 71, 102–110.

Ganster, D.C. and Rosen, C.C. (2013) Work stress and employee health: A multidisciplinary review. *Journal of Management*, 39(5), 1085–1117.

Ganster, D.C. and Schaubroeck, J. (1991) Work stress and employee health. *Journal of Management*, 17(2), 235–271.

Garfield, J. (1995) Social stress and medical ideology. *Stress and Survival*, 3, 111–134.

George, J.M., Reed, T.F., Ballard, K.A., Colin, J. and Fielding, J. (1993) Contact with AIDS patients as a source of work-related distress: Effects of organizational and social support. *Academy of Management Journal*, 36, 157–171.

Glass, D. (1977) *Behavior Patterns, Stress and Coronary Disease*. Hillsdale, NJ: Erlbaum.

Glisson, C.A. and Martin, P.Y. (1980) Productivity and efficiency in human service organizations as related to structure, size and age. *Academy of Management Journal*, 23(1), 21–37.

Gmelch, W.H. (1982) *Beyond Stress to Effective Management*. New York: Wiley.

Goldenhar, L.M., Williams, L.J. and Swanson, N.G. (2003) Modelling relationships between job stressors and injury and near-miss outcomes for construction labourers. *Work and Stress*, 17, 218–241.

Greenberg, J. (1999) *Managing Behavior in Organizations*. Upper Saddle River, NJ: Prentice Hall.

Greenhaus, J.H. and Parasuraman, S. (1987) A work–nonwork interactive perspective of stress and its consequences. In J.M. Ivancevich and D.C. Ganster (eds), *Job stress: From Theory to Suggestion*, 37–60. New York: Haworth.

Gryna, F.M. (2004) *Work Overload! Redesigning Jobs to Minimize Stress and Burnout*. New York: ASQ Quality Press.

Hagihara, A., Tarumi, K., Miller, A.S. and Morimoto, K. (1997) Type A and type B behaviors, work stressors, and social support at work. *Preventive Medicine*, 26(4), 486–494.

Handy, C. (1985) *Understanding Organizations*. London: Penguin.

Hayes, C.T. and Weathington, B.L. (2007) Optimism, stress, life satisfaction and job burnout in restaurant managers. *Journal of Psychology*, 141, 565–579.

Health and Safety Executive (2007) *An Analysis of the Prevalence and Distribution of Stress in the Construction Industry*. London: Crown (RR518).

Hinze, J. (1988) Safety on large building constructions projects. *Journal of Construction of Engineering Management*, 114(2), 286–293.

Holmes, T.H. and Rahe, R.H. (1967) The Social Readjustment Rating Scale. *Journal of Psychosomatic Research*, 11, 213–218.

Ibem, E.O., Anosike, M.N., Azuh, D.E. and Mosaku, T.O. (2011) Work stress among professionals in the building construction industry in Nigeria. *Australasian Journal of Construction Economics and Building*, 11(3), 46–57.

Ilgen, D.R. andHollenbeck, J.R. (1991) The structure of work: Job design and roles. In M.D. Dunnette and L.M. Hough (eds), *Handbook of Industrial and Organizational Psychology*, 2, 165–207. Palo Alto, CA: Consulting Psychologists Press.

Ismail, A.M., Thomson, M.J., Vergara, G.V., Rahman, M.A., Singh, R.K. and Gregorio, G.B., (2010) Designing resilient rice varieties for coastal deltas using modern breeding tools. In C.T. Hoanh, B.W. Szuster, K.S. Pheng, A.M. Ismail and A.D. Nobel (eds), *Tropical Deltas and Coastal Zones: Food Production, Communities and Environment at the Land-Water Interface*, 154–165. Wallingford: CAB.

Jackson, J. (1989) *A Description of the Objective and Subjective Dimensions of Daily Activity Patterns of 15 Adolescents with Disabilities*. Unpublished paper funded by a grant from the American Occupational Therapy Foundation, Rockville.

Jamal, M. (1990) Relationship of job stress and type-A behavior to employees' job satisfaction, organizational commitment, psychosomatic health problems and turnover motivation. *Human Relations*, 43, 727–738.

Janssen, P.P.M., Bakker, A.B. and de Jong, A. (2001) A test and refinement of the demand–control–support model in the construction industry. *International Journal of Stress Management*, 8(4), 315–332.

John, G. and Martin, J. (1984) Effects of organizational structure on marketing-planning credibility and utilization of plan output. *Journal of Marketing Research*, 21, 170–183.

Joiner, T.A. (2001) The influence of national culture and organizational culture alignment on job stress and performance: Evidence from Greece. *Journal of Managerial Psychology*, 16(3), 229–242.

Kahn, R.L. and Byosiere, P. (1992) Stress in organizations. In M.D. Dunnette and L.M. Hough (eds), *Handbook of Industrial and Organizational Psychology*. Palo Alto, CA: Consulting Psychologists Press, 571–650.

Kahn, R.L., Wolfe, D.M., Quinn, R.M., Snowek, J.D. andRosenthal, R.A. (1964) *Organizational Stress: Studies in Role Conflict and Ambiguity*. Chichester: Wiley.

Kasl, S. V. (1992) Surveillance of psychological disorders in the workplace. In G. Keita and S. Sauter (eds), *Work and Well-Being: An Agenda for the 1990s*. Washington, DC: American Psychological Association.

Katz, D. and Kahn, R.L. (1978) *The Social Psychology of Organizations*. New York: Wiley.

Keenan, T., Copper, C.L. and Marshall, J. (1980) *Stress and the Professional Engineer*. Chichester: Wiley.

Klitzman, S. and Stellman, J.M. (1989) The impact of the physical environment on the psychological well-being of office workers. *Social Science and Medicine*, 29(6), 733–742.

Kohli, A.K. and Jaworski, B.J. (1990) Market orientation: The construct, research propositions and managerial implications. *Journal of Marketing*, 54, 1–18.

Kraimer, M.L. and Wayne, S.J. (2004) An examination of POS as a multidimensional construct in the context of an expatriate assignment. *Journal of Management*, 30, 209–237.

Kuper, H., Singh-Manoux, A., Siegrist, J. and Marmot, M. (2002) When reciprocity fails: Effort–reward imbalance in relation to coronary heart disease and health functioning within the Whitehall II study. *Occupational and Environmental Medicine*, 59, 777–784.

Lacaille, D., Sheps, S., Spinelli, J.J., Chalmers, A. and Esdaile, J.M. (2004) Identification of modifiable work-related factors that influence the risk of work disability in rheumatoid arthritis. *Arthritis Rheum*, 51, 843–852.

Landy, F.J. (1992) Work design and stress. In G. Keita and S. Sauter (eds), *Work and Well-Being: An Agenda for the 1990s*. Washington, DC: American Psychological Association.

Lazarus, R.S. and Folkman, S. (1984) *Stress, Coping and Adaptation*. New York: Springer-Verlag.Leung, M.Y. (2004) An international study on the stress of estimators. *The Hong Kong Surveyor*, 15(1), 49–52.

Leung, M.Y., Chan, I.Y.S. and Yu, J.Y. (2012) Preventing construction worker injury incidents through the management of personal stress and organizational stressors. *Journal of Accident Analysis and Prevention*, 48, 156–166.

Leung, M.Y., Ng, S.T., Skitmore, M. and Cheung, S.O. (2005) Critical stressors influencing construction estimators in Hong Kong. *Construction Management and Economics*, 23(1), 33–43.

Leung, M.Y., Skitmore, M. and Chan, Y.S. (2007) Subjective and objective stress in construction cost estimation. *Construction Management and Economics*, 25(10), 1063–1075.

Leung, M.Y., Sham, J. and Chan, Y.S. (2007) Adjusting stressors – Job-demand stress in preventing rustout/burnout in estimators. *Surveying and Built Environment*, 18(1), 17–26.

Leung, M.Y., Chan, Y.S., Chong, A. and Sham, J.F.C. (2008) Developing structural integrated stressors-stress models for clients' and contractors' cost engineers. *Journal of Construction Engineering and Management*, 134(8), 635–643.

Leung, M.Y., Chan, Y.S. and Olomolaiye, P. (2008) Impact of stress on the performance of construction project managers. *Journal of Construction Engineering and Management*, 134(8), 644–652.

Leung, M.Y., Zhang, H. and Skitmore, M. (2008) Effects of organisational support for construction cost engineer stress. *Journal of Construction Engineering and Management*, 134(2), 83–93.

Leung, M.Y., Chan, Y.S. and Yu, J.Y. (2009) Integrated model for the stressors and stresses of construction project managers. *Journal of Construction Engineering and Management*, 135(2), 126–134.

Leung, M.Y., Chan, Y.S. and Chong, A.M.L. (2010) Chinese values and stressors of construction personnel in Hong Kong. *Journal of Construction Engineering and Management*, 136(12), 1289–1298.

Leung, M.Y., Chan, Y.S. and Yuen, K.W. (2010) Impacts of stressors and stress on the injury incidents of construction workers in Hong Kong. *Journal of Construction Engineering and Management*, 136(10), 1093–1103.

Leung, M.Y., Chan, Y.S. and Yu, J.Y. (2011) Preventing construction worker injury incidents through the management of personal stress and organizational stressors. *Accident Analysis and Prevention*, 156–166.

Lewy, A.J., Kern, H.E., Rosenthal, N.E. and Wehr, T.A. (1982) Bright artificial light treatment of a manic-depressive patient with a seasonal mood cycle. *American Journal of Psychiatry*, 139, 1496–1498.

Lim, L.J.W. and Ling, F.Y.Y. (2012) Human resource practices of contractors that lead to job satisfaction of professional staff. *Engineering, Construction and Architectural Management*, 19(1), 101–118.

Lingard, H. (2004) Work and family sources of burnout in the Australian engineering profession: A comparison of respondents in dual and single earner couples, parents and non-parents. *Journal of Construction Engineering and Management*, 130(2), 290–298.

Lingard, H. and Francis, V. (2005) Does work–family conflict mediate the relationship between job schedule demands and burnout in male construction personnel and managers? *Construction Management and Economics*, 23, 733–745.

Lingard, H. and Francis, V. (2006) Does a supportive work environment moderate the relationship between work-family conflict and burnout among construction professionals? *Construction Management and Economics*, 24(2), 185–96.

Loosemore, M. and Waters, T. (2004) Gender differences in occupational stress among professionals in the construction industry. *Journal of Management in Engineering*, 20 (3), 126–132.

Love, P.E.D., Haynes, N.S. and Irani, Z. (2001) Construction managers' expectations and observations of graduates. *Journal of Managerial Psychology*, 16, 579–593.

Lundberg, U. and Cooper, C.L. (2010) *The Science of Occupational Health*. Oxford: Wiley-Blackwell.

MacDavitt, K., Chou, S.S. and Stone, P.W. (2007) Organizational climate and health care outcomes. *Joint Commission Journal on Quality and Patient Safety*, 33(1), 45–56.

Mansfield, N.R. and Odeh, N.S. (1991) Issues affecting motivation on construction projects. *International Journal of Project Management*, 9 (2), 93–98.

Margolis, B.L., Kroes, W.H. and Quinn, R.P. (1974) Job stress – An unlisted occupational hazard. *Journal of Occupational Medicine*, 16(10), 659–661.

Maslach, C. and Jackson, S.E. (1984) Burnout in organizational settings. In S. Oskamp (ed.), *Applied Social Psychology Annual: Applications in Organizational Settings*, 5, 133–153. Beverly Hills, CA: Sage.

Maslach, C., Schaufeli, W.B. and Leiter, M.P. (2001) Job burnout. *Annual Review of Psychology*, 52, 397–422.

Maslanka, H. (1996) Burnout, social support and AIDS volunteers, *AIDS Care*, 8(2), 195–206.

Maslow, A.H. (1954) *Motivation and Personality*. New York: Harper.

Mayo, M., Sanchez, J.I., Pastor, J.C. and Rodriguez, A. (2012) Supervisor and coworker support: A source congruence approach to buffering role conflict and physical stressors. *International Journal of Human Resource Management*, 23(18), 3872–3889.

Miller, D. (1987) The structural environmental correlates of business strategy. *Strategic Management Journal*, 8(1), 55–76.

Mintzberg, H. (1979) *The Structuring of Organizations*. Englewood Cliffs, NJ: Prentice-Hall.

Mmaduakonam, A.E. (1997) *Occupational Stress Counselling for Workers' Survival*. Enugu: Academic.

Morioka, I., Miyai, N. and Miyashita, K. (2006) Hot environment and health problems of outdoor workers at a construction site. *Industrial Health*, 44(3), 474–480.

Mostert, K., Peeters, M. and Rost, I. (2011) Work–home interference and the relationship with job characteristics and well-being: A South African study among employees in the construction industry. *Stress and Health*, 27(3), e238–e251.

Mustapha, F.H. and Langford, D. (1990) *What skills do effective site managers bring to their work?* Proceedings of the CIB W-90 Conference.

Nevels, P. (1986) *Autonomy and stress. Harvard Business Review*, 64(3), 164.

Niedhammer, I., Chastang J.F., David, S., Barouhiel, L. and Barrandon, G. (2006) Psychosocial work environment and mental health: Job–strain and effort–reward imbalance models in a context of major organizational changes. *International Journal of Occupation Environment and Health*, 12, 111–119.

Ochieng, E.G. and Price, A.D.F. (2010). Managing cross-cultural communication in multicultural construction project teams: The case of Kenya and UK. *International Journal of Project Management*, 28(5), 449–460.

O'Driscoll, M.P. and Beehr, T.A. (1994) Supervisor behaviors, role stressors and uncertainty as predictors of personal outcomes for subordinates. *Journal of Organizational Behaviour*, 15, 141–155.

Oldham, G.R. and Rotchford, N.L. (1983) Relationships between office characteristics and employee reactions: A study of the physical environment. *Administrative Science Quarterly*, 28, 542–556.

Organ, D.W. (1979) The meaning of stress. *Business Horizons*, 22 (3), 32–40.

Petersen J.S. and Zwerling C. (1998) Comparisons of health outcomes among older construction and blue-collar employees in the United States. *American Journal of Industrial Medicine*, 34(3), 280–287.

Pheng, L.S. and Chuan, Q.T. (2006) Environmental factors and work performance of project managers in the construction industry. *International Journal of Project Management*, 24, 24–37.

Porter, L.W. and Lawler, E.E. (1968) *Managerial Attitudes and Performance*. Homewood, IL: Richard D. Irwin.

Quick, J.C. and Quick, J.D. (1984) *Organizational Stress and Preventative Management*. New York: McGraw-Hill.

Quick, J. C. and Quick, J. D. (1997) *Preventive Stress Management in Organizations*. Washington, DC: American Psychological Association.

Quinn, R., Seashore, S., Kahn, R., Mangion, T., Campbell, D., Staines, G. and McCullough, M. (1971) *Survey of Working Conditions: Final Report on Univariate and Bivariate Tables*. Washington, DC: US Government Printing Office, Document No. 2916–0001.

Rhoades, L. and Eisenberger, R. (2002) Perceived organizational support: A review of the literature. *Journal of Applied Psychology*, 87, 698–714.

Rizzo, J.R., House, R.J. and Lirtzman, S.I. (1970)Role conflict and ambiguity in complex organizations. *Administrative Science Quarterly*, 15(2), 150–164.

Rogers, R.W. (1983) Cognitive and physiological processes in fear appeals and attitude change: A revised theory of protection motivation. In J. Cacioppo and R. Petty (ed.), *Social Psychophysiology*, 153–176. New York: Guilford Press.

Robbins, T.L. and DeNisi, A.S. (1998) Mood vs. interpersonal affect: Identifying process and rating distortions in performance appraisal. *Journal of Business and Psychology*, 12(3), 313–325.

Rowley, C. and Jackson, K. (2011) *Human Resource Management: The Key Concepts*. Oxford: Routledge.

Sauter, S. and Murphy, L. R. (1995) *Organizational Risk Factors for Job Stress*. Washington, DC: American Psychological Association.

Scheier, M.F. and Carver, C.S. (1985) Optimism, coping and health: assessment and implications of generalized outcome expectancies. *Health Psychology*, 4, 219–247.

Schloesser, R.J., Lehmann, M., Martinowich, K., Manji, H.K. and Herkenham, M. (2010) Environmental enrichment requires adult neurogenesis to facilitate the recovery from psychosocial stress. *Molecular Psychiatry*, 15, 1152–1163.

Segars, A.H., Grover, V. and Teng, J.T. (1998) Strategic information systems planning: Planning systems dimensions, internal coalignment and implications for planning effectiveness. *Decision Sciences*, 29(2), 303–345.

Selmer, J. and Fenner, C.R. (2009) Spillover effects between work and nonwork adjustment among public sector expatriates. *Personnel Review*, 38(4), 366–379.

Seltzer, J., Numberof, R.E. and Bass, B.M. (1989) Transformational leadership: Is it a source of more burnout and stress? *Journal of Health and Human Resources Administration*, 12, 174–185.

Selye, H. (1956) *The Stress of Life*. New York: McGraw-Hill.

Siefert, K., Jayaratne, S. and Chess, W.A. (1991) Job satisfaction, burnout and turnover in health care social workers. *Health and Social Work*, 16(3), 193–202.

Smith, L.A., Roman, A., Dollard, M.F., Winefield, A.H. and Siegrist, J. (2005) Effort–reward imbalance at work: the effects of work stress on anger and cardiovascular disease symptoms in a community sample. *Stress and Health: Journal of International Society for the Investigation of Stress*, 21(2), 113–128.

Smithers, G.L. and Walker, D.H.T. (2000) The effect of the workplace on motivation and demotivation of construction personnel. *Construction Management and Economics*, 18(7), 833–841.

Sogaard, A.J., Dalgard, O.S., Holme, I., Roysamb, E. and Haheim, L.L. (2007) Associations between Type A behavior pattern and psychological distress. *Social Psychiatry and Psychiatric Epidemiology*, 43(3), 216–223.

Stattin, M. and Jarvholm, B. (2005) Occupation, work environment and disability pension: A prospective study of construction workers. *Scandinavian Journal of Public Health*, 33, 84–90.

Sunindijo, R. and Zou, P. (2013) Conceptualizing safety management in construction projects. *Journal of Construction Engineering and Management*, 139(9), 1144–1153.

Sawacha, E., Naoum, S. and Fong, D. (1999) Factors affecting safety performance on construction sites. *International Journal of Project Management*, 17(5), 309–315.

Sutherland, V.J. and Davidson, M.J. (1993) Using a stress audit: The construction site manager experience in the UK. *Work and Stress*, 7(3), 273–286.

Swanson, V., Power, K.G. and Simpson, R.J. (1998) Occupational stress and family life: A comparison of male and female doctors. *Journal of Occupational and Organizational Psychology*, 71, 237–260.

Takeuchi, R., Shay, J.P. and Li, J.T. (2008) When does decision autonomy increase expatriates' adjustment? An empirical test. *Academy of Management Journal*, 51, 45–60.

The Standard (2007) *Call to Alter Work Hours to Reduce Heatstroke Risk*. Retrieved from http://www.thestandard.com.hk/news_print.asp?art_id = 49957& sid = 14665065

Varma, A., Denisi, A.S. and Peters, L.H. (1996) Interpersonal affect and performance appraisal: A field study. *Personnel Psychology*, 49, 341–360.

Vigoda, E. (2000) Internal politics in public administration systems: An empirical examination of its relationship with job congruence, organizational citizenship behavior and in-role performance. *Public Personnel Management*, 29, 185–210.

Vischer, J.C. (2007) The effects of the physical environment on job performance: Towards a theoretical model of workspace stress. *Stress and Health*, 23, 175–184.

Yarnold, P.R. and Grimm, L.G. (1982) Conformity, interpersonal dominance and the Type A personality. Proceedings of 3rd Annual *Meeting of the Society of Behavioral Medicine*, Illinois University, Chicago, 1–13.

Yip, B. (2008) Professional efficacy among building professionals. *The Conference for Building Professionals*. Gold Coast, Australia, 24–26.

Yip, B. and Rowlinson, S. (2009) Job burnout among construction engineers working within consulting and contracting organizations. *Journal of Management in Engineering*, 25(3), 122–130.

Wittchen, M., Krimmel, A., Kohler, M. and Hertel, G. (2013) The two sides of competition: Competition-induced effort and affect during intergroup versus interindividual competition. *British Journal of Psychology*, 104(3), 320–338.

Zaleznik, A., Kets de Vries, M.F.R. and Howard, J. (1977) Stress reactions in organizations, syndromes, causes and consequences. *Behavioral Science*, 22, 151–162.

Zou, P.X.W., Zhang, G. and Wang, J. (2007) Understanding the key risks in construction projects in China. *International Journal of Project Management*, 25(6), 601–614.

5

Consequences of Stress

5.1   Consequences of Stress Affecting Construction Personnel

The consequences of occupational stress for various aspects of performance in different disciplines have received increasing research attention over the past two decades (see e.g. Quick et al. 1997 – for general employees; Pflanz and Ogle 2006 – for military personnel; Scott et al. 2006 – for nurses; Leung et al. 2005; Leung et al. 2006; Leung et al. 2008; Leung et al. 2010 – for construction personnel). It is clear that stress affects construction personnel in various ways. For example, one of the key roles of a project manager is to make decisions. However, when under excessive stress, such decisions might become rigid, simplistic and superficial (Cherrington 1994). Based on the extensive literature on stress management, the consequences can be categorised in terms of three aspects: (inter)personal, task and organisational performance (Black and Gregersen 1990; Leung et al. 2005; Leung et al. 2006; Leung, Chan and Olomolaiye 2008; Leung, Chan and Yuen 2010).

5.2   (Inter)Personal Performance of Construction Personnel

5.2.1   Personal Satisfaction

Satisfaction can be defined as a pleasurable or positive emotional state resulting from the satisfactory appraisal of experiences in one's personal and working life (Locke 1976). It is a key attitudinal variable affecting employee outcomes such as job performance and turnover intention. There are generally two types of satisfaction: cognitive and affective (Moorman 1993).

Stress Management in the Construction Industry, First Edition.
Mei-yung Leung, Isabelle Yee Shan Chan and Cary L. Cooper.
© 2015 John Wiley & Sons, Ltd. Published 2015 by John Wiley & Sons, Ltd.

Affective job satisfaction refers to the degree to which an individual is satisfied with subjective feelings (happiness, pleasure and so on) relating to aspects of their jobs, such as working hours, salary, benefits and environment (Leung et al. 2002). Cognitive job satisfaction denotes an individual's reaction to the overall work situation which results from a comparison between actual and desired consequences (Leung et al., 2005; Mathieu and Zajac 1990). Jobs not only offer people extrinsic value, but also intrinsic satisfaction which enables one to realise a passion and achieve self-actualisation. Cognitive job satisfaction can, in turn, influence affective satisfaction. An individual under stress tends to be less satisfied at work.

Studies have confirmed that the higher the level of stress, the lower the job satisfaction of managers, engineers (Maurya 2012), social professions (Trzcieniecka-Green et al. 2012), workers (Cooper and Starbuck 2005) and even students (Yang and Moon 2011). Although there are inconsistencies within this body of work in terms of the effects of job satisfaction on the performance and commitment of employees (for instance, Shore and Martin (1989) show that a high level of job satisfaction leads to a high level of commitment, while Iaffaldano and Muchinsky (1985) and Locke (1976) find a negative relationship), research in the construction industry context indicates that greater satisfaction can result in better performance (Henne and Locke 1985; Leung et al. 2004; Leung, Chen and Yu 2008). In fact, job satisfaction, performance and commitment are interrelated (Leung, Chan and Dongyu 2011). Construction personnel have an emotional attachment to their tasks and a will to contribute their professional knowledge to get the job done. Accordingly, this can lead to improvements in their performance and, ultimately, their job satisfaction.

5.2.2   Interpersonal Relationships

Construction projects often involve multiple stakeholders and interorganisational project team members, such as client representatives, designers, consultants, contractors and subcontractors. Communication and cooperation between these parties is essential for effective information processing in a project, which directly influences success (Galegher, Kraut and Egido 1990).

Individuals suffering from stress may have a higher chance of encountering communication problems, such as misunderstandings or even conflict (Robbins and Fray 1980), which affect the quality of both individual and group decision making. If individuals are distracted by stressors, this may affect their ability to focus in the present moment (that is, the act of communication itself), making it difficult to concentrate fully and listen to what their counterparts are saying. This can easily result in misunderstandings. Any conflict arising in the design and construction stages of a project may cause substantial delays and financial losses.

On the other hand, stress also affects patience (Semmer 2003), which may result in the individual not really focusing on what a colleague is saying, but instead interrupting to express his or her own point of view. People under stress may also respond defensively (Cameron et al. 2005), making it hard for them to see others' points and respond with empathy. Stress influences the working relationships between individuals and their colleagues, supervisors and subordinates. This may induce a general lack of concern for colleagues, or disrespect for, distrust of and/or dislike for them (Defrank and Cooper 1987; Holt 1993; Leung et al. 2005; Leung et al. 2006; Leung, Chan and Olomolaiye 2008). It is therefore important for construction personnel to manage their stress levels.

5.3   Task Performance of Construction Personnel

5.3.1   Project Outcomes

The evidence confirms there is a close relationship between stress and task performance for construction personnel. Task performance in this context denotes making correct decisions, meeting client requirements, planning an effective schedule and controlling project duration and costs (Haynes and Love 2004; Leung et al. 2005; Leung et al. 2006; Leung, Chan and Olomolaiye2008). As Selye (1982) shows, people under stress often perform at higher levels initially, but if the stress continues and exhaustion sets in, a range of problems ensue. A decline in productivity is one of the most obvious task consequences of excessive stress (Leung, Chan and Olomolaiye 2008; Tarafdar et al. 2007). Construction personnel are the cornerstone of any project. Their productivity and performance directly influence success in various areas, such as time, cost and quality. For instance, the lowered productivity of a stressed project manager may lead him or her to make mistakes in programme planning, influencing the entire process and duration of the project and leading to increased costs (Murray and Zagaretos 2001). Similarly, a decline in effectiveness for a quantity surveyor may delay the estimation process and result in errors, influencing the planning of other parties and causing loss to the project owners (Leung et al. 2005). Where architects are concerned, stress may influence their understanding and expression of the client's requirements, ultimately undermining satisfaction with the project outcomes.

Understanding and meeting the requirements of the client has been shown to be one of the key determinants of project performance (Ling, Ibbs and Hoo 2006). When suffering from stress, an individual will be able to allocate fewer resources to the performance of tasks as a result of being consumed by the stressful events; in other words, his or her effectiveness will decrease. Even though an individual may be able to spend the same amount of time on a task when under stress, his or her concentration levels will be reduced as a result and productivity will also decrease significantly (i.e., lowered efficiency) (Krueger 1989; Robertson and Cooper 2012). There is a longstanding body of work on the impact of stress on the productivity of various types of personnel such as teachers (Blix et al. 1994) and nurses (Milliken, Clements and Tillman 2007). However, studies of the stress of construction personnel, and in particular its impact on their productivity, are still rare.

5.3.2   Safety Behaviours

Construction workers are the key, indeed indispensable, contributors to every project. They have a direct and significant impact on the success of construction projects and on the profitability of construction companies (Applebaum 1999). However, the dynamic nature of the construction industry and the demanding nature of their jobs mean that construction workers not only need to have huge amounts of physical vigour and energy, but are also forced to work uncertain though long hours. They may sometimes need to work on evenings, weekends and holidays to finish a job or to respond to an emergency, especially if the task needs special government approval because it causes a public nuisance (US Department of Labor 2007). Such conditions can be frustrating and emotionally taxing, affecting construction workers' energy levels and attention span. Stress also affects awareness of, and compliance with, safety measures. Failure to be aware of and comply with such measures is the major cause of occupational accidents and injuries (Mearns et al. 2001). Construction injury incidents are very costly and have a significant impact on progress. It is therefore important to reduce the injury risk to construction workers by managing their stress.

5.4   Organisational Performance of Construction Personnel

5.4.1   Sense of Belonging

To a certain extent, stress can influence organisational effectiveness, which is particularly important to construction stakeholders. Organisational ineffectiveness simply refers to the financial impact resulting from the poor performance of employees (McGrath 1976), such as withdrawal behaviour and a reduced sense of belonging to the organisation (Leung, Chan and Olomolaiye 2008; Nandram and Klandermans 1993; Schuler 1982).

Sense of belonging denotes the degree to which an employee feels accepted, respected and supported by others in the organisation (Baumeister and Leary 1995). It is a fundamental human need (Maslow 1943). Construction personnel may suffer from depression if they lack a sense of belonging, which can have a detrimental impact on both them and the organisation. A strong relationship has been demonstrated between a sense of belonging and stress symptoms such as depression and anxiety (Cockshaw and Shochet 2010).

5.4.2   Intention to Stay

In addition, turnover rate has been recognised as another direct concern for organisations (Cox 1993). The costs of staff turnover can result in significant financial losses in terms of training, induction, the uncertainty of recruiting new staff and so on (Chartered Institute of Personnel and

Development 2013). The overall annual cost of turnover due to stress and mental health problems could be as high as £1.35 billion in the UK. In addition to meeting the individual's physical and material needs, jobs also provide, to a certain extent, social ties which enable one to connect with others and demonstrate a social identity (Brunetto and Farr-Wharton 2002). For these reasons, it is not easy for an individual to give up work. However, excessive stress reduces the employee's intention to stay (Gray-Toft and Anderson 1981; Jamal 1999; Lee and Shin 2005). To prevent such loss of personnel, it is recommended that construction stakeholders provide their staff with adequate organisational support through means such as health and stress management seminars or workshops. Such efforts may not only prevent employees from suffering from distress, but also enhance their intention to stay.

5.5   Development of a Conceptual Model of Stress and Performance

5.5.1   Conceptual Model of Stress and Performance

The extensive review of the literature outlined above has identified three major aspects of performance for construction personnel: (inter)personal, task and organisational. These factors, in turn, can be decomposed into personal satisfaction, interpersonal relationships, project outcomes, safety behaviours, sense of belonging and intention to stay. Too much work, or tasks which are too complicated (i.e., work stress), induce worry, anxiety (emotional stress) and headaches (physical stress). These directly influence personal satisfaction, relationships with colleagues, productivity, observance of safety behaviours, sense of belonging and intention to stay. Figure 5.1 illustrates the relationships between the three types of stress (work, emotional and physical) and their impact on performance for construction personnel.

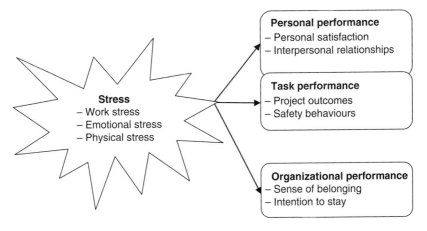

Figure 5.1   Conceptual model of stress and performance.

5.5.2   Relationship between Stress and Performance

Is stress necessarily a bad thing? There is no definitive answer and indeed the relationship between stress and performance for individuals varies from study to study. According to a meta-analysis of empirical studies, 46% demonstrate a *negative linear* relationship, 13% a *positive linear* relationship and only 4% a *curvilinear* relationship (Muse, Harris and Feild 2003). These inconsistent results may be due to the different theoretical frameworks, measurement methods, connotations and conditions of stress used in the studies reviewed (Muse, Harris and Feild 2003). In this section, various theories supporting the three major possible relationships between stress and performance are discussed.

5.5.2.1 *Linear relationships*

Muse, Harris and Feild (2003) demonstrate statistically that stress has, in general, been perceived as detrimental to individual performance (a *negative linear* relationship; see Figure 5.2). This is because performance is impaired due to the deprivation of time, energy and attention required to cope with stress (Jamal 1984). According to Vroom (1964), stress can impair individual performance by inducing involuntary physical responses and narrowing individual perceptions at work. The negative linear relationship between stress and performance is supported by various studies (Beehr, Walsh and Taber 1976; Friend 1982; Jamal 1984, 2007; Westman and Eden 1991).

On the other hand, Meglino (1977) links stress to challenge and suggests that the relationship between performance and stress can actually be

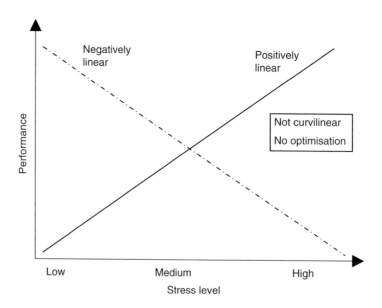

Figure 5.2   The linear relationships between stress and performance (e.g., Friend 1982; Meglino 1977).

positively linear for simple tasks (see Figure 5.2). According to his model, individuals under low stress do not perceive a challenge and are therefore less likely to make the effort to enhance their performance. Those at medium levels of stress will be moderately aroused by challenge and perform at a moderate level. Lastly, those encountering high levels of stress will be able to perform at an optimal level because they are motivated by the optimum degree of challenge. Other studies confirm this positive relationship for employees in general (Cohen 1980; Hatton et al. 1995) and construction personnel (Leung, Chan and Olomolaiye 2008).

5.5.2.2 Curvilinear relationships

Stress is not necessarily bad (Selye 1974). Yerkes and Dodson (1908) were the first to identify the scientific principle of the inverted U-shaped relationship between stress and performance. As illustrated in Figure 5.3, too much or too little stress can lead to impaired performance, while moderate levels can produce optimum performance (Yerks and Dodson 1908). In fact, individuals who do not have enough stress may suffer from understimulation (known as rustout) as they are not fully activated; equally, however, if they encounter too much stress, they may suffer from overstimulation (burnout) because the amount of effort they have to invest in coping might divert their efforts from the job (Gmelch 1982). Hence, only moderate levels of stress will act as a motivator for individuals to work well and deliver optimum performance, because such employees are sufficiently activated while still able to devote adequate amounts of energy to the job (Gmelch 1982). Research confirms the curvilinear relationship between stress and performance for employees in general (Anderson, Leu and Kant 1988; Duffy 1962; Selye 1976; Zaccaro and Riley 1987) and construction personnel in

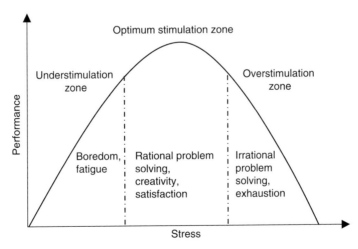

Figure 5.3 The curvilinear relationship between performance and stress (e.g., Anderson 1976; Gmelch 1982; Leung et al. 2005; Yerkes and Dodson 1908).

particular (Leung 2004; Leung et al. 2005; Leung et al. 2006; Leung, Chan and Olomolaiye 2008).

The optimum level of stimulation is not fixed and may vary across individuals and tasks. It has been proposed that different tasks require different levels of arousal (Meglino 1977; Selye 1976; Zajonc 1965). For example, difficult or intellectually demanding tasks may require a lower level of arousal to facilitate concentration and deliver optimal performance. However, tasks demanding stamina or persistence may be performed better at higher levels of arousal, which increase motivation. Therefore, a more precise model can be proposed (see Figure 5.4). However, it can also be argued that the change in direction after the optimum point, where high stress causes cognitive narrowing and rigidity of behaviour, is particularly applicable to complex tasks where novel responses, problem solving or attention to many elements are necessary.

Although Djebarni (1996) identifies an inverted U-shaped relationship between job stress and leadership effectiveness for site managers in Algeria, limited research has been undertaken on the link between stress and performance for construction personnel specifically. While the task, personal and organisational performance of construction personnel are key to the success of every project, this is clearly affected by stress (work, emotional or physical). Leung et al. (2005, 2008) have undertaken comprehensive research in the construction field which confirms the close relationships, both linear and curvilinear, between the three kinds of stress and the three types of performance. Based on the above, the following stress–performance models are developed specifically for construction personnel which incorporate both linear and curvilinear relationships. In other words, stress is not necessarily bad and its relationships with types of performance are not necessarily negative and linear. The inverted U-shaped model demonstrates that there is an optimisation zone of stress which facilitates the highest levels of performance for each individual.

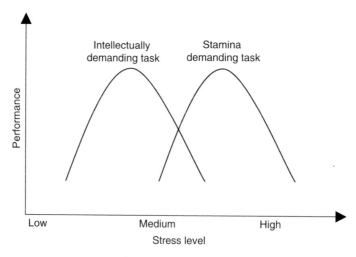

Figure 5.4 The relationship between performance and stress for different tasks (e.g., Meglino 1977; Selyes 1976; Zajonc 1965).

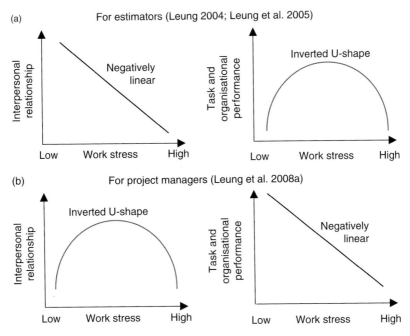

Figure 5.5 The relationships between stress and performance for estimators and project managers.

All in all, these studies demonstrate the relationships between stress and three main types of performance, namely personal, task and organisational performance. Figure 5.5 summarises the various issues in each category for estimators and project managers. Our previous work indicates there are both linear and nonlinear relationships between stress and performance for these groups. The relationship between work stress and interpersonal relationships for estimators is negatively linear (Leung 2004; Leung et al. 2005; see Figure 5.5a), while for project managers is an inverted U-shape (Leung et al. 2008; see Figure 5.5b). The relationship between work stress and both task and organisational performance for estimators is an inverted U-shape, while a negative relationship is found for project managers. It can be seen that it is impossible to identify a single or universal stress–performance relationship across industrial or professional boundaries. These relationships are highly dependent on the individual, the profession, the type of stress encountered, the relevant aspect of performance and so on.

The differences in the impact of stress on various types of personnel may be due to the different interactions they experience between stress, the nature of their work and the support received, given their different positions in projects and organisations (Bowen, Edwards and Lingard 2013). For instance, the negative relationship between work stress and task performance for construction project managers may indicate that their stress level has already passed the optimisation point and so will result in poor task performance. On the other hand, since the role of project managers requires them to remain closely connected with other team members, they may be trained to have strong skills in building and maintaining relationships. Certain types of work stress may mean they are required to undertake specific tasks involving coordinating and connecting with others. Accordingly, only excessive stress would worsen their interpersonal relationships (i.e, there is an inverted U-shaped relationship between work stress and interpersonal relationships). In contrast, interpersonal skills trainings for estimators may be comparatively lesser, meaning that work stress may easily worsen their relationships (in other words, there is a negatively linear relationship between work stress and interpersonal relationships for this group). Notwithstanding the difference in the impact of stress on the performance of various kinds of construction personnel, the relationship between work stress and overall project performance is also an inverted U-shape (Leung et al. 2005).

5.6   Research Results on Stress and Performance of Construction Personnel

In order to further explore and understand the impact of stress on performance among various types of construction personnel, we conducted both statistical and case studies.

5.6.1   Statistical Studies

The three main types of performance cover a total of six specific types: personal (comprising personal satisfaction and interpersonal relationships); task (comprising project outcomes and safety behaviours); and organisational performance (comprising sense of belonging and intention to stay). To measure the performance of construction personnel, we adopted 15 items from previous studies covering all types (Leung et al. 2004; Leung et al. 2006; Leung, Chan and Olomolaiye 2008; Leung, Chen and Yu 2008; Leung, Chan and Yuen 2010; Leung et al. 2011). Respondents were asked to indicate their agreement to the various items using a 7-point Likert-type scale ranging from 1 (strongly disagree) to 7 (strongly agree). The data were then analysed using SPSS 20.0.

Due to the different sample sizes, we calculated the means and standard deviations for all dimensions of performance. As shown in Table 5.1 and Figure 5.6, the highest mean score was obtained for interpersonal relationships (M = 5.070, SD = 0.900) and the lowest for safety behaviours (an outcome that was rated by construction workers only: M = 3.747, SD = 1.550). Personal satisfaction, project outcomes, intention to stay and sense of belonging were scored second, third, fourth and fifth, respectively. A reliability analysis was performed to test the internal consistency of each performance factor. The alpha values of the items measuring the six types of performance were all greater than 0.5, indicating that they were sufficiently reliable (Hair et al., 1998).

Table 5.1 Performances of various construction personnel.

Performance	Sample size	Mean	SD	Alpha
Personal performance				
P1 Personal satisfaction	81	4.592	.665	N/A.
+ I am satisfied with my overall performance				
P2 Interpersonal relationships	406	5.070	.900	.577
+ I am satisfied with the relationships between my colleagues and me				
+ I get along well with others at work				
+ I trust the people who work with me				
Task performance				
P3 Project outcomes	497	4.550	.950	.749
+ I seldom make wrong decisions				
+ I can meet clients' requirements easily				
+ Most of the schedules I plan are effective				
P4 Safety behaviours	274	3.747	1.550	.862
− I ignore safety regulations to get the job done				
− I break work procedures				
− I bend safety rules to achieve production targets				
Organisational performance				
P5 Sense of belonging	171	4.180	1.507	.877
+ I am proud to tell others that I am part of this organisation				
− I feel like an outsider in the organisation				
+ I generally feel that people accept me in the organisation				
P6 Intention to stay	402	4.550	1.102	.582
− I frequently think that the company is not suitable for me				
+ I will accept any kind of job assignment so as to keep the organisation working				
− I intend to leave this company				

Note: +/- denotes items with aligned or opposite direction with the indicating stressor (for '-' items, scores were reversed om data analyses)

The construction industry emphasises teamwork, since most projects involve multiple stakeholders. These results indicate that two types of personal performance, interpersonal relationships and personal satisfaction, scored higher than others. Organisational performance, comprising a sense of belonging and intention to stay, received a mean score far lower than that of personal performance. This may reflect the nature of the project-based construction sector, which is characterised by low levels of commitment, a lack of a sense of belonging and weak intention to stay (Spatz 2000). Task performance, consisting of project outcomes and safety behaviours, was rated lowest. Due to the unique and complex nature of construction projects, tasks are normally difficult for personnel to complete and it is not uncommon for them to even fail to do so. Moreover, construction workers are placed at the lowest level of the organisation and are often required to finish huge amounts of work within a limited time. Given such

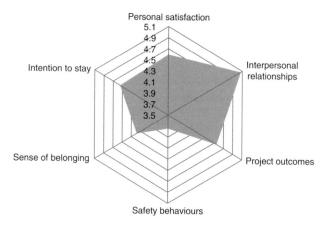

Figure 5.6 Performance reported by construction personnel.

an overemphasis on delivery, they may be forced to enhance their productivity at the expense of safety (Mullen 2004).

5.6.1.1 *Comparison of the performance of construction personnel by age*

A one-way between-groups analysis of variance (ANOVA) was performed to investigate the impact of age on the performance scores of construction personnel. Respondents were divided into four groups (Group 1: 29 or under; Group 2: 30–39; Group 3: 40–49; Group 4: 50 and above). Of the sample, 26.3% were aged 29 or under, 25.8% were 30–39, 31.9% were 40–49 and 15.7% were 50 or above. Group 1 reported the lowest of all mean scores in four performance categories (interpersonal relationships: $M=4.933$, $SD=0.944$; project outcomes: $M=4.375$, $SD=0.948$: sense of belonging $=3.859$, $SD=1.315$; intention to stay: $M=4.235$, $SD=1.143$), while groups 2, 3 and 4 scored the highest in interpersonal relationships ($M=5.263$, $SD=0.915$), project outcomes ($M=4.660$, $SD=0.920$) and intention to stay ($M=4.711$, $SD=1.156$), respectively. Compared with the other five types of performance, safety behaviours were scored relatively low across all age groups with means ranging from 3.486 to 3.983. Across all six types of performance, only two demonstrated statistically significant differences between age groups at the $p<0.05$ level, namely interpersonal relationships ($F=2.790$, $p=0.040$) and intention to stay ($F=3.103$, $p=0.027$) (see Table 5.2 and Figure 5.7).

To analyse the age differences in every type of performance between different age groups, post hoc comparisons were conducted using the Tukey HSD test (see Table 5.2). There was a significant difference in the scores for interpersonal relationships and intention to stay between groups 1 and 2. The relationships score for group 1 ($M=4.933$, $SD=0.944$) was significantly lower than for group 2 ($M=5.263$, $SD=0.915$) with a mean difference of -0.330 ($p=0.045$). The intention to stay score of group 1 was

Table 5.2 One-way between-groups ANOVA for performance of construction personnel in different age groups.

Performance	Age Group	Mean	SD	F (ANOVA)	Sig. (ANOVA)	Sig. (Levene)	Group	Mean Diff.	SE	Sig.
							Post hoc test for performance with significant diff. scores			
P1 Personal satisfaction	≤ 29	4.933	.770	.483	.695	.733				
	30–39	4.490	.558							
	40–49	4.650	.666							
	≥ 50	4.516	.841							
P2 Interpersonal relationships	**≤ 29**	**4.933**	**.944**	**2.790**	**.040**	**.645**	**30–39**	**−.330**	**.126**	**.045**
	30–39	5.263	.915							
	40–49	5.005	.831							
	≥ 50	5.159	.917							
P3 Project outcomes	≤ 29	4.375	.948	2.410	.066	.385				
	30–39	4.609	.921							
	40–49	4.660	.920							
	≥ 50	4.558	1.015							
P4 Safety behaviours	≤ 29	3.690	1.340	1.161	.325	.021				
	30–39	3.486	1.434							
	40–49	3.852	1.644							
	≥ 50	3.983	1.815							
P5 Sense of belonging	≤ 29	3.859	1.315	1.045	.374	.024				
	30–39	4.362	1.232							
	40–49	4.313	1.761							
	≥ 50	3.981	1.386							
P6 Intention to stay	**≤ 29**	**4.235**	**1.143**	**3.103**	**.027**	**.554**	**30–39**	**−.464**	**.169**	**.031**
	30–39	4.699	1.029							
	40–49	4.529	1.091							
	≥ 50	4.711	1.156							

Note: **Bolded items** - significant between-group differences

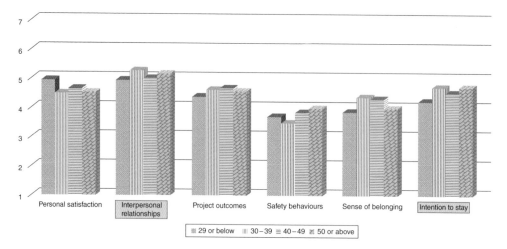

Figure 5.7 Performance reported by construction personnel in different age groups
Note: Highlighted items - Significant differences revealed in the one-way ANOVA (refer to Table 5.2).

significantly lower than for group 2 with a mean difference of –0.464
(p = 0.031). No significant differences were found across age groups in
terms of personal satisfaction, project outcomes, safety behaviours and
sense of belonging.

5.6.1.2 *Comparison of performance of construction personnel by gender*

The ANOVA results also indicated significant differences in project outcome
scores between male and female personnel (see Table 5.3 and Figure 5.8). In
the male-dominated construction industry, male respondents scored project
outcomes significantly higher than females, with a mean difference of 0.429
(p = 0.030). There were no other significant gender differences in the other
five types of performance.

5.6.2   Correlation Analysis

To investigate the relationships between the three kinds of stress (work,
emotional and physical) and the six types of performance (personal satisfaction,
interpersonal relationships, project outcomes, safety behaviours, sense of
belonging and intention to stay), a Pearson correlation analysis was conducted.
To further investigate the inverted U-shaped relationship between stress and
performance, the square of stress was also included in the correlation analysis.
The inter-correlations among this sample of construction personnel are shown
in Table 5.4.

The results were as follows: (i) personal satisfaction (P1) was significantly
positively correlated with interpersonal relationships (P2: 0.725, p < 0.01),
project outcomes (P3: 0.615, p < 0.01), sense of belonging (P5: 0.562, p < 0.01)
and intention to stay (P6: 0.673, p < 0.01); (ii) interpersonal relationships

Table 5.3   One-way between-groups ANOVA for performance of construction personnel by gender.

Performance		Gender	Mean	SD	F (ANOVA)	Sig. (ANOVA)	Sig. (Levene)
P1	Personal satisfaction	Male	4.589	.662	0.071	.790	.404
		Female	4.717	1.076			
P2	Interpersonal relationships	Male	5.085	.891	1.689	.195	.372
		Female	4.817	1.057			
P3	**Project outcomes**	**Male**	**4.610**	**.953**	**4.745**	**.030**	**.006**
		Female	**4.181**	**.667**			
P4	Safety behaviours	Male	3.732	1.544	1.192	.276	.673
		Female	4.583	1.873			
P5	Sense of belonging	Male	4.184	1.518	.145	.704	.617
		Female	3.944	1.272			
P6	Intention to stay	Male	4.532	1.101	.858	.355	.894
		Female	4.772	1.095			

Note: **Bolded items** - significant between-group differences

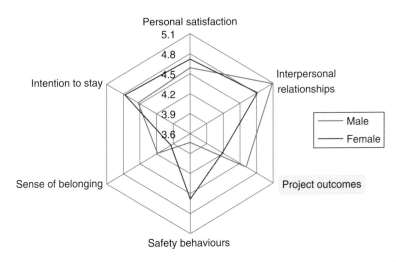

Figure 5.8   Performance reported by male and female construction personnel. Note: Highlighted item- Significant differences revealed in the one-way ANOVA (refer to Table 5.3).

(P2) was significantly positively correlated with project outcomes (P3: 0.548, $p < 0.01$) and intention to stay (P6: 0.428, $p < 0.01$), but negatively with safety behaviours (P4: –0.286, $p < 0.01$); (iii) project outcomes (P3) was significantly positively related to turnover intention (P6: 0.334, $p < 0.01$) and negatively to safety behaviours (P4: –0.203, $p < 0.01$); (iv) safety behaviours (P4) was significantly negatively correlated with sense of belonging (P5: –0.410, $p < 0.01$)

Table 5.4 Pearson correlation analysis between stress and performance for construction personnel.

Performance	Stress			Stress²			Performance					
	WS	ES	PS	WS²	ES²	PS²	P1	P2	P3	P4	P5	P6
P1. Personal satisfaction	-.326**	-.023	-.226*	-.368**	-.009	-.226*	1.000					
P2. Interpersonal relationships	-.201**	-.309**	-.260**	-.018	-.260**	-.259**	.725**	1.000				
P3. Project outcomes	-.247**	-.234**	-.122*	.024	-.163**	-.103	.615**	.548**	1.000			
P4. Safety behaviours	.127*	.080	-.064	-.165*	.045	-.058	—	-.286**	-.203**	1.000		
P5. Sense of belonging	.103	.131	.079	.001	-.156*	.080	.562**	.167	-.125	-.410**	1.000	
P6. Intention to stay	-.037	-.360**	-.152*	-.023	-.310**	-.152*	.673**	.428**	.334**	-.228**	.681**	1.000

Note: WS – work stress, ES – emotional stress, PS – physical stress.

*= correlation is significant at the 0.05 level (two-tailed).

**= correlation is significant at the 0.01 level (two-tailed).

and intention to stay (P6: –0.228, p < 0.01); and (v) sense of belonging (P5) was significantly positively correlated to intention to stay (P6: 0.681, p < 0.01).

All three kinds of stress were negatively correlated with interpersonal relationships (P2) and project outcomes (P3); that is, work stress (S1: –0.201 and –0.247, p < 0.01), emotional stress (S2: –0.309 and –0.234, p < 0.01) and physical stress (S3: –0.260, p < 0.01 and –0.122, p < 0.05). Intention to stay was significantly negatively correlated with emotional (S2: –0.360, p < 0.01) and physical stress (S3: –0.152, p < 0.05), while personal satisfaction had a significantly negative relationship with work (S1: –0.326, p < 0.01) and physical stress (S3: –0.226, p < 0.05). In addition, safety behaviours was positively correlated with work stress (S1: 0.127, p < 0.05). The results also indicated the presence of inverted U-shaped relationships between stress and performance. Emotional stress was linked in this way with four types of performance, namely interpersonal relationships (P2: –0.260, p < 0.01), project outcomes (P3: –0.163, p < 0.01), sense of belonging (P5: 0.156, p < 0.05) and intention to stay (P6: –0.310, p < 0.01). Physical stress was related to personal satisfaction (P1: –0.226, p < 0.05), interpersonal relationships (P2: –0.259, p < 0.01) and intention to stay (P6: –0.152, p < 0.05). Finally, work stress was linked to personal satisfaction (P1: –0.368, p < 0.05) and safety behaviours (P4: –0.165, p < 0.05).

5.7   Discussion

5.7.1   Performance of Construction Personnel by Age

The study has demonstrated significant differences in terms of interpersonal relationships and intention to stay between construction personnel in different age groups, while it also shows that no age-related differences were found for the other four aspects of performance (personal satisfaction, project outcomes, safety behaviours and sense of belonging) (see Table 5.2). A significant difference in scores for *interpersonal relationships* was found between groups 1 (≤29) and 2 (30–39). Construction projects are normally characterised by complicated teams of multiple stakeholders and intra- or intergroup conflict in the development and construction process is common (e.g., the government might expect the job to start immediately, suppliers might not deliver materials on time, residents might try to stop work due to socio-environmental issues). It may be difficult for younger construction personnel who are at the start of their career and have very limited experience to manage this complexity well (Keenan and Newton 1987). On the other hand, construction personnel aged 30–39 are expected to have become skilled and proficient. Unlike their older compatriots (aged 40 or above), who are the key people in the project team, they have normally acquired the skills required to communicate effectively with various stakeholders and to ensure they allocate sufficient time to maintaining good working relationships.

Intention to stay for construction personnel aged 29 or below was significantly lower than for those aged 30–39. As newcomers to a construction organisation, younger personnel need to learn about industrial values, organisational culture, expected behaviour and social knowledge in order to prepare themselves to understand and participate as organisational members (O'Reilly, Chatman and Caldwell 1991). Younger personnel may well have less intention to stay during their first few years with an organisation, since they will be unfamiliar with the system and culture in terms of aspects such as the organizational structure, and career development prospect (O'Reilly, Chatman and Caldwell 1991; Reichers, Wanous and Steele 1994). On the other hand, construction personnel in the 30–39 bracket have got through their 'green' period and will be familiar with their job, their colleagues and the organisational structure. They will normally be capable not only of handling frontline tasks but also dealing with complicated relationships within and outside the project. Most of them are likely to have been promoted to middle management, which may enhance their intention to stay with the organisation for their long-term career development. In view of the different situations experienced by the two age groups, the differences in their scores for intention to stay are readily understandable.

5.7.2   Performance of Construction Personnel by Gender

As Table 5.3 shows, there were no significant gender differences in terms of personal satisfaction, interpersonal relationships, safety behaviours, sense of belonging and turnover intention. However, male construction personnel reported significantly better *project outcomes* than their female counterparts. Unlike other occupations such as teaching and nursing, the tasks which require to be performed in construction industry roles can be physically demanding (e.g., frequent site visits in often extreme outdoor conditions). Men tend to have the advantage in terms of the ability to endure such a hostile and potentially threatening environment (Dainty, Bagilhole and Neale 2000), which may explain this finding. For instance, in order to achieve each milestone in the construction process, meet the client's requirements and handle unexpected events, construction personnel often have to work overtime. However, many women play dual roles, as employees but also wives and/or mothers. Due to their biological and social circumstances, female staff may have comparatively fewer resources and social support to draw on in order to tackle their work, in terms of both effort and time, which may result in their poorer self-reported performance on project outcomes.

5.7.3   Stress and Performance

The results indicated significant inverted U-shaped relationships between work stress and two types of performance (personal satisfaction and safety behaviours); between emotional stress and four types (interpersonal relationships, project outcomes, sense of belonging and intention to stay); and physical stress and three types (personal satisfaction, interpersonal relationships and intention to stay).

Due to the characteristics of construction projects (such as their temporary and unique nature), personnel must deal with many difficult tasks.

Work stress emphasises the mismatch between expected and actual ability to perform a job task. Tasks which are too complicated or simple and work-load which is too much or too little, can both cause work stress (Cox 1993; Leung et al. 2009). Neither too much nor too little work stress is favourable. Only a moderate level can lead to good performance. Project managers usually take a leader's role, handling various dynamic and challenging tasks. If these demands exceed their abilities, personal dissatisfaction is likely to result. On the other hand, excessive amounts of routine administration tasks may be too simple and boring for construction personnel, given their abilities. This can also result in low levels of personal satisfaction, as they may perceive themselves as being underutilised.

Construction workers are on the frontline and are exposed to various hazards and risks onsite. Too much work stress will force them to focus exclusively on productivity at the expense of safety through means such as taking unsafe shortcuts to complete tasks as soon as possible (Loosemore and Waters 2004). However, too little work stress may induce rustout. In such a situation, workers may consider it unnecessary to pay attention to safety because the tasks assigned to them are too easy to pose any challenges or risks (Leung et al. 2012).

Individuals experiencing positive emotions (such as happiness and relaxation) are more likely to be able to concentrate on their work and approach it with a clear mindset (Bishop et al. 2004; Jain and Kobti 2011) and hence to perform the task as efficiently as expected (Robins and Judge 2003). However, construction personnel with high levels of *emotional stress* are likely to show symptoms such as depression, frustration, loss of temper, fatigue or anxiety. This leads directly to sluggishness at work and negatively affects their interaction with subordinates, colleagues and supervisors. This may also lead to errors such as submitting incorrect design plans or documents or producing poor quality work in general. Such a situation ultimately induces burnout and a reduced sense of belonging and intention to stay with the company (Cropanzano, Rupp and Byrne 2003; Ducharme, Knudsen and Roman 2007; Wright and Cropanzano 1998). For example, a quantity surveyor suffering from emotional stress is likely to find it harder to concentrate on estimation work and hence may risk arriving at an inaccurate result. This, in turn, is likely to be accompanied by blame from their supervisor or the client, resulting in a weak sense of belonging and intention to stay.

On the other hand, too little emotional stress is also associated with poor performance among construction personnel. When an individual is suffering from a certain level of work stress, he or she may also experience associated emotional stress, such as worry about achieving tasks. However, if suffering from understimulation, a lack of worry or emotional arousal may be symptoms of poor performance due to a lack of motivation. For instance, a junior quantity surveyor who has been assigned only simple and repetitive tasks is unlikely to suffer from worry or frustration due to work stress. However, he or she may still lack the motivation to perform tasks and cooperate with others, resulting in poor relationships with colleagues, a reduced sense of belonging and perhaps even an intention to leave the organisation.

The results also indicated the presence of inverted U-shaped relationships between physical stress and performances of construction personnel, which implies that either too little or too much of the former impairs the latter. Physical symptoms may make it hard for construction personnel to keep their minds on their work and maintain interest and satisfaction. Working in a stressful environment with too much physical stress not only reduces personal satisfaction, but also affects communication with others and damages relationships. In addition, too much physical stress may also reduce the intention of construction personnel to stay with their organisation, because withdrawing from work would constitute an escape from the source of stress and hence reduce it.

On the other hand, too little physical stress may, to some extent, indicate that the individual is being understimulated. Physical stress symptoms, such as headaches and musculoskeletal pain, are very common among workers in fast-paced metropolitan cities. These symptoms are likely to worsen if an individual is suffering from excessive workload. In contrast, someone who is understimulated may not experience any physical stress due to the lack of motivation. This, again, is likely to result in low levels of personal satisfaction and a lack of desire to cooperate with others, resulting in poor relationships. Construction personnel in this position may feel that remaining with their current organisation will not benefit their career development; if so, they are likely to look for a new position where they can develop themselves and achieve self-actualisation. However, given that physical stress is prevalent in the construction industry and that prolonged suffering from physical stress can lead to serious health problems and even mortality in extreme cases, construction personnel should make an attempt to reduce it.

5.7.4   Interrelationships among the Six Types of Performance

The results have also revealed interrelationships among the six types of performance, namely personal satisfaction, interpersonal relationships, project outcomes, safety behaviours, sense of belonging and intention to stay. It is more likely that staff who have good relationships with others and enjoy the harmonious and friendly atmosphere which results from such team spirit will also express personal satisfaction, have a stronger sense of belonging, be more likely to intend to stay in the organisation and perform better in tasks, all of which can result in better project outcomes (Yang 2009). Others may have poor or even conflictual communication with colleagues. Such employees are less likely to seek support from others or form and participate well in teams, resulting in poor quality work and potentially unfinished or delayed tasks (Beehr et al. 2000). As such personnel may be easily discouraged and lacking in job satisfaction, they may engage less with their work, resulting in poorer organisational performance in term of a reduced sense of belonging (Coomber and Barriball 2007).

It is interesting to note that safety behaviours are negatively correlated with the other four types of performance. In this study, only construction workers rated the items on safety. Construction site safety measures are necessary, not only because of laws and regulations, but also because incidents can result in delay to the construction process which has cost implications. However, complying with safety regulations may mean that construction workers cannot operate in the way they are accustomed to and have to carry out additional procedures involving specific items of personal protection equipment. This, to a certain extent, may affect their efficiency as well as their productivity in term of project outcomes (Choudhry and Fang, 2008). It is not uncommon to find construction workers being caught and penalised by safety officers, because they often ignore safety measures in favour of convenience and efficiency (Lombardi et al. 2009). The strict safety regulation required by an organisation may made construction workers feel uncomfortable and frustrated, which could result in a reduced sense of belonging and a greater likelihood of intention to leave. Strict safety working procedures can also interfere with their traditional approaches within work teams, leading to problems communicating with colleagues, particularly safety officers.

5.8   Case Studies

5.8.1   A Mega International Airport Project

As part of a huge airport development project, this project aimed to construct a structural reinforced concrete frame for the terminal building. The work was to be carried out by a large contractor for a contract sum of HK$150 million within a period of 330 days. Due to the tight programme for the airport development overall, the project of structural framework had to be completed within a specified timescale.

Since problems were being experienced, primarily insufficient labour and an inconvenient site location, the airport authority permitted the use of migrant labour to increase staffing levels and complete the work on time. In addition, many workers were required to live onsite to minimise travel time and enhance communication. Due to the tight timescale, the project manager only went home twice a week, while the workers living onsite returned home every Saturday and came back to work on Monday. This arrangement was beneficial not only in solving traffic problems but also increasing the amount of staffing hours available to the project.

The most important task in the construction process was to manage the vast number of staff and deliver large amounts of materials to the site. Sufficient staffing was critical to meeting the project completion date, but involved complicated personnel management processes. In the middle of the process, the project manager became aware that interpersonal relationships between the local and migrant labourers were poor. Local workers complained that the migrants did not follow proper working procedures. On the other hand, the migrants felt that they were working hard but their efforts were not appreciated by the locals.

In addition, to ensure work sequences were efficiently scheduled, all the site facilities were provided within a limited timeframe. One of the workers complained he had had to work harder and more quickly because the mobile crane or movable steel scaffolds were no longer provided once the rental time was up. However, the sequence could not be completely controlled by the contractor. The project team members were concerned about the frequent and sudden special orders which made the schedule even tighter.

Due to the complexity of the airport project, site safety was important for all construction personnel involved in the work. The project manager reported that unsafe behaviours onsite were common, particularly among the migrant workers. When he found out that some of them had not brought a safety helmet or gloves, he had no option but to prevent them entering the working area until they had found and were wearing suitable protective clothing.

In order to establish a reliable control system, the project manager had set up a cooperation centre onsite to support all parties to work together. The architect advised that the operation of the centre had to be in line with the company's policy, rules and procedures. Although the project manager had worked hard to set up the centre as a platform for monitoring the construction process, he was confused by his boss's decision to reduce the number of workers allocated to the project because of the high labour costs.

The scheduling of this aspect of the airport project was tight, causing stress for the personnel involved. As a result, they expressed understandable dissatisfaction with their jobs (see Table 5.5). The project manager and construction workers felt tired and unsatisfied because of the tight timescale and long working hours onsite. The project manager said that this was 'obviously not a preferred working style'. One of the engineers also indicated that it was difficult for him to control working sequences and project progress because of the frequent unexpected situations arising onsite. Because this led to him being asked regularly to reschedule and reorient the project plan, he felt that his performance was unsatisfactory and his productivity had been reduced.

The relationships between different parties were also affected by the scheduling and resultant stress levels. To complete the project on time, a large number of migrant workers had been hired. However, the project manager found it difficult to deal with the conflicts between local and migrant labourers, inducing high levels of stress and conflict with his team. The local workers complained that they 'couldn't get along with the migrants, because they didn't follow the correct working sequences and exposed us to high risks'. Moreover, other professionals such as architects and engineers also reported difficulties working together. The engineer thought that the changes made to the drawings had disturbed the original plan, which subsequently influenced the whole progress of the project, while the architect indicated that the other team members had hesitated to amend the original drawings. With all of them suffering from different kinds of stressors and stress, team members did not work well together.

Stress also influenced project outcomes in terms of the effectiveness of scheduling, project quality and ability to meet the client's requirements. The project manager said that the short duration of the project made him feel stressed. He tried to solve problems directly so he could go home early, which would have resulted in greater effectiveness and better

Table 5.5 Performance of various construction personnel under stress.

Performance	Construction personnel
Personal satisfaction	PM: Because of the tight project timescale, long working hours and long travelling hours, I feel that I am always very tense. This is obviously **not a preferred** working style for me. Eng: The project programme is always being rescheduled, which results in a great deal of stress for me. There are often unexpected situations onsite and it is challenging for me to respond on time. Being honest, my **productivity has been low.** CW: Due to the long working hours, I am stressed and my personal life is messed up. I really **hate this situation,** but I have to make a living.
Interpersonal relationships	PM: We have employed many migrant labourers to provide sufficient staffing for the project. However, it is difficult for me to manage both migrant and local workers, which causes me huge amounts of stress and has resulted in frequent **quarrels with my team members**. Arch: The drawings have been changed several times, so my working hours are continually being extended. It is difficult for me to **work with** the project manager and engineers to revise the drawings. Eng: Our team made a good job of planning the working sequences in consideration of the short project duration. However, we have often received special orders, which have disrupted our original plan. It is hard for me to **communicate with other project members**, especially when I am under stress. CW: There are many migrant workers in this project. However, I **can't get along with** them, because they don't follow the correct working sequences and expose us to high risks.
Project outcomes	PM: The short duration of the project has made me stressed. When I feel stressed, I try to solve the problem directly to manage it. I hope to **solve work problems** quickly and go home early, which would result in **greater effectiveness and better outcomes**. Eng: When I am affected by stress, I might focus only on meeting the deadline, neglecting **quality and the client's requirements.** QS: We have to work long hours and make many calculations of the salary of the migrant workers and the cost of different working sequences. It is easy to get the **calculations and estimates wrong** when I am under so much stress.
Safety behaviours	PM: **Unsafe behaviours** are more common among migrant workers, especially in such a stressful project with a tight schedule. I often stopped migrant workers when I saw that they were not wearing safety helmets. People are only allowed to enter the project site if they were wearing **safety equipment**. CW: The migrant workers are more likely to **ignore safety precautions** and **break some safety rules**. This is especially serious when the project completion stage is approaching and the schedule is really tight and everybody is **stressed**. I have to pay attention to them and **avoid unnecessary accidents**.

(Continued)

Table 5.5   (*Continued*)

Performance	Construction personnel
Sense of belonging	PM: Although we established an organisational structure onsite, my boss still influenced ordinary decisions. Sometimes I cannot understand my boss's decisions, such as reducing staffing levels because of high labour costs. He might not have realised that the project timescale is too tight and the amount of staff not enough. My professional judgement has not been respected . . . To be honest, the **organisational culture really conflicts** with my views. Arch: To facilitate cooperation between various parties, the project manager set up a cooperation centre onsite. However, it does not operate in line with our company's rules and procedures, so I have had to follow two sets of working procedures, which has exacerbated my stress levels. I have reported this situation to my company. However, no feedback was received, as usual . . . I sometimes think that **my company does not value its employees.** CW: I was brought into the project on a short-term contract and will return to my home country after completion. It is hard and unnecessary for me to **foster a sense of belonging** to this company, especially on such a stressful project.
Intention to stay	PM: I really **want to quit my current job** because of the long working hours and the significant stress. Eng: Because of my huge workload and long working hours onsite, I have had some health problems, such as headaches and sleep disorders. I have sometimes thought **of leaving this project**. In fact I was thinking about it last night because my job is just too stressful. CW: The long working hours have given me backache and musculoskeletal pain. Hence, I am **thinking of finding a new job**.

Note: PM = Project manager; Arch = Architect; Eng = Engineer; QS = Quantity surveyor; CW = Construction worker.

outcomes. However, the quantity surveyor and engineer felt differently. The former had experienced significant stress arising from the amount of different calculations he had had to perform to ascertain the cost of bringing in migrant labour and approaching the work in different sequences. This had impaired his performance and caused him to make incorrect calculations. Similarly, the engineer also indicated that he tended to focus on meeting deadlines rather than ensuring quality when he felt under pressure.

These personnel also mentioned their experiences and observation of onsite safety behaviours and performance under stress. The project manager had noticed that unsafe behaviours were common among the migrant workers and he had often had to restrict their entry to the site if they were not properly equipped. The construction worker also complained that the migrant staff were 'more likely to ignore safety precautions', and felt that he and his local colleagues had had to pay attention to the migrant workers in order to avoid accidents. The poor safety behaviours of one group of workers on a construction site inevitably increases both the workload and stress level of other personnel, affecting their performance in turn.

Mention was also made of organisational performance issues. The project manager had established an onsite cooperation centre, but he still needed to adhere to his boss's decisions. He was unable to understand some of the choices made by his manager such as reducing the staffing allocated to the project. He also felt disrespected because his boss had not taken his professional judgement into account. This made it very difficult for him to lead the project team and he came to the conclusion that the 'organisational culture conflicted with my values'. On the other hand, the architect also mentioned the cooperation centre, but highlighted that it had not been established exactly in line with the company's rules. Hence, the architect felt forced to follow two sets of procedures. He tried to seek support from his company but none was forthcoming. This caused him stress and he realised that the organisational culture had come into conflict with his own values. One of the migrant workers also said that it had been hard for him to foster a sense of belonging to the company, because he was on a short-term contract and would return to his home country once the work had been completed.

Stress also reduced the intention to stay with the current organisation among this group of workers. The project manager said that he wanted to quit because of the long working hours and the high stress. The engineer also reported physical health problems due to the substantial workload. He had 'sometimes thought of leaving the project'. The construction worker complained about life onsite and working overtime, which had caused him backache and musculoskeletal pain. He was also thinking about finding a new job.

5.8.2 A Project on a Constrained Site at Traffic Congestion Zone Hong Kong Island

This project aimed to build a high-class 28-storey residential building in a city centre. It was led by a private developer and managed by its in-house project team. The construction was carried out by a large contractor for a contract sum of HK$110 million within a period of 290 days. As the project was located close to the city centre, the working area was constrained and space was limited, causing site management issues.

Having a limited working space significantly influenced the project sequence, creating difficulties for the construction process. The project manager had had to revise the site planning, construction sequences and programming and arrange an offsite bending yard for steel fabrication. Moreover, since the workers were also restricted in terms of space, they were required to do more multiple-stage tasks. For example, excavated soil could not be removed from the site directly at the digging points and had to be moved to a waste collecting point for onward transportation; also, windows and doors needed to be placed in temporary storage and then moved to the required block of the construction, indirectly if necessary. Some workers said that they had had to work harder and take more time to complete such multiple-stage tasks.

As well as having limited space, the short duration of the project also caused stress. The project manager complained that the client would often try to speed up progress and he had needed to rearrange working sequences accordingly. Some tasks, such as the structural framework, window installation and plastering, had to be carried out at almost the same time. Hence, the project manager had had to revise the plan and arrange for the steel bending yard.

The construction workers pointed out that a poor and crowded site environment with waste materials lying around, such as steel bars and other debris, increased the possibility of incidents. It was unsafe for them to work long hours in such an environment. However, the project manager had also asked his workers to put in longer hours to complete the building on time. The engineers expressed an opposing opinion and pointed out that workers may neglect unsafe conditions when they are dealing with multiple tasks and working regular overtime. In fact, the company had received a warning from the Labour Department about the poor site environment. In order to improve onsite safety, the project team provided sufficient light for all site accesses and distributed safety equipment to all workers. In order to reduce the risk of incidents, the project manager and site foremen would not allow workers onsite without a safety helmet or gloves.

The subcontractors reported increased labour costs because their workers had had to handle multiple tasks during the construction process. The architect also worried about subcontractors employing unskilled workers because their budget did not provide for sufficient qualified staff. The restricted project environment also constrained the amount of workers provided by the subcontractors. One worker said that his boss had not provided him with enough colleagues, as would be required to finish the project. There was also congestion of workers in the same area of the construction site. Indeed, one complained that the area was so crowded that his own work was disturbed by that of others. The project manager said that he would try to arrange the site facilities in order to avoid such interference. The project as a whole took place in a stressful atmosphere. One of the workers admitted that he was exhausted due to his heavy workload while the project manager wanted to take a short leave so that he could rest.

Due to the site constraints and tight schedule, the whole project took place in a stressful atmosphere. One of the workers admitted that he was dissatisfied due to the stress levels involved, while the project manager also reported high stress resulting in low productivity and dissatisfaction (see Table 5.6). Both the project manager and the engineer complained about the revision of the site plan because of the limited workspace. The changes increased the workload of the engineer, who felt let down by what he saw as poor site management. This affected the performance of both professionals. More importantly, it is apparent that they were also dissatisfied with their jobs and level of output.

Table 5.6   Performance of various construction personnel under stress.

Performance	Construction personnel
Personal satisfaction	PM: I often work overtime to revise site planning and working sequences, which is stressful. However, my **productivity is low due to the stress** and I am **not satisfied with my current job**. Eng: The project manager often changes site planning, which increases my workload. I am really **disappointed in and dissatisfied** with the poor site management. CW: The site is too small. There is not enough space for all the workers who are needed, so I have to handle multiple tasks at the same time. I am **dissatisfied due to the stressful nature of the work**.
Interpersonal relationships	PM: One of the subcontractors wanted to increase their labour expenses, because their workers handle multiple tasks. To be honest, I cannot accept their claim, which causes me stress and has even led to me losing my temper. It is **difficult for me to maintain good relationships with them**. Eng: I usually speak very *fast* and *work* in a *panic* when I am under stress . . . It has been difficult for me to **communicate with** the project manager about the site planning. CW: Due to the constrained site environment, we often work alongside people in other trades, which has definitely impaired our progress and extended our working hours. Hence, we often **fight with each other** about the work sequences onsite.
Project outcomes	PM: The difficulties with this project have caused me stress. But I directly s**olve the problems and complete my tasks effectively**. Eng: My supervisor and client blamed me for the **poor efficiency and quality** of my work when I was stressed. Because of my poor performance, the drawings cannot be completed on time, which influences the **whole project progress**, in turn induces more stress. QS: I do feel stressed due to the frequent changes of site management and increase in construction costs and labour expenses. But when I am under stress, I normally p**erform more proficiently and seldom make mistakes**.
Safety behaviours	PM: We were warned by the Labour Department about our poor site environment. **Workers who are stressed tend to ignore safety requirements**. To improve safety, I forbade workers who were not wearing safety helmet to enter the project site. Eng: Workers need to handle multiple tasks due to the limited space and insufficient labour available. This makes them more stressed and causes them to **neglect unsafe conditions** onsite. CW: The site environment is too crowded with waste materials. It's frustrating to work in such a poor and unsafe site environment for a long time. I try to finish my job early by any means, even sometimes **taking shortcuts and breaching safety procedures**.

(Continued)

Table 5.6   *(Continued)*

Performance	Construction personnel
Sense of belonging	Eng: I have experienced vast amounts of stress on this job. The project manager has often asked me to work overtime and spoken to me about my performance. My **sense of belonging** to this organisation is affected when I am under stress. QS: The project manager seems not to place much emphasis on the cost estimation and project budget. I feel that it will be really hard for me to **achieve my career goals** in this company.
Intention to stay	Arch: The construction project team lacks the ability to construct my design. So I have been asked to revise my drawings many times. If I feel it is stressful for me to continue working on the current project, there is **no doubt that I would like to leave**. QS: I feel bored in my current position. The project manager does not take us seriously. I am **thinking of changing jobs**. CW: I have had no time to rest, which really increases my stress levels. So I am looking for other job opportunities. I want to **quit my current job** because of work overload.

Note: PM = Project manager; Arch = Architect; Eng = Engineer; QS = Quantity surveyor; CW = Construction worker.

Many parties were involved in the residential building project. Several personnel mentioned the challenges of maintaining good relationships. The project manager explained that he had been unable to authorise the subcontractors' claim to increase their salary budget, which had caused him stress and influenced his relationship with subcontractors. The engineer had experienced a great deal of stress as a result of his heavy workload. He said that as a result he spoke too quickly and felt panicky, which made it hard for him to communicate with the project manager. Moreover, workers from different trades had had to work together due to the cramped site environment. Inevitably, they disturbed each other, leading to longer working hours. One construction worker said that he had often fought with colleagues about who would do what piece of work in what sequence.

Construction personnel mentioned several different project outcomes resulting from their high levels of stress. The project manager admitted that he faced many challenges and accordingly felt stressed. However, he had tackled the problems directly, resulting in better outcomes. One of the quantity surveyors also mentioned that he had felt stressed due to the frequent changes in site management and increased construction costs. However, he thought that this level of pressure had actually improved his performance. In contrast, the engineer reported feeling stressed when he was blamed by the project manager for his inefficiency. This resulted in poor performance and late submission of drawings, which influenced the project as a whole.

This project was located in a city centre with limited working space, leading to a poor safety environment and a warning from the Labour Department about the insufficient provision of safety equipment. The project manager observed that workers under stress tend to ignore safety behaviours and explained that he had taken various actions to improve safety. The engineer was concerned about the multiple tasks carried out by workers, which distracted them and might result in incidents, especially during rush periods when they were stressed by the tight timeframe. Moreover, one of the construction workers complained about the crowded site environment and the amount of waste materials onsite. Such a poor site environment is definitely not favourable to frontline workers; he expressed his frustration, explaining that he would 'finish his job early by any means, even sometimes taking shortcuts and breaching safety procedures'.

The stress experienced by construction personnel also affected their sense of belonging to the organisation and the project. The engineer was stressed because he was often asked to work overtime and then blamed for his poor performance. He said that his sense of belonging had been affected as a result. One quantity surveyor mentioned that the project manager did not place much importance on cost estimation, which made him feel that it would be hard for him to achieve his life and career goals with this company.

As a result of this stress, the intention of these construction personnel to stay with the company was also reduced. The architect complained that the project manager had asked him to revise his drawings many times. He felt stressed and made it clear that he 'would like to leave the company'. As the project manager did not take the quantity surveyor seriously, he was bored with his position and had thought about changing jobs. Moreover, the construction workers mentioned that he had no time to rest and had considered quitting as a result of this work overload.

5.9   Practical Implications

Construction personnel aged 29 or below tended to have the lowest score on interpersonal relationships and intention to stay among the six types of performance considered in this study. Hence, it is necessary to take action to improve the performance of younger construction personnel. A *training scheme* taking a *mentoring approach*, such as *Learning the Ropes* (Louis 1980) or the *Engineering Graduate Scheme A Training* (Hong Kong Institute of Engineers 2013) may help personnel new to the industry develop realistic expectations of their job requirements, working environments and social norms in the organisation (Feldman 1976). As a result, they may feel better able to cope with work stress by seeking instrumental support and thus enhancing various aspects of their performance. In the meantime, they can also learn how to understand the organisational culture and adjust their working attitudes accordingly (Louis 1980). Moreover, a supportive atmosphere can also give progressive opportunities to newcomers working independently, reduce their vulnerability in difficult situations and equip them with the professional competencies necessary to deliver project outcomes (Barnes 2009). Such improved performance can further enhance their motivation to work on complicated construction projects, which can ultimately help to foster a sense of belonging to the company (Lam, Lo and Chan 2002).

To optimise the ability to cope with emotional stress, *social gatherings* for construction personnel should also be encouraged, to help people build good relationships with colleagues and establish a harmonious working environment. Social networks are the key basis for instrumental and emotional support seeking, which are both essential in stress management (Mccoll, Lei and Skinner 1995). On the other hand, given that physical and mental health are the main concern of stress management in the construction industry, *medical and psychological consultants* should also be involved in providing services and advice to individuals and professional consultancy to the company (e.g., planning and conducting occasional social activities, holding physical health seminars) (Forman 1981).

As well as this, senior management should also *review the workload and expectations of each participant at regular intervals.* Construction projects often involve dynamic and uncertain situations that directly affect the workload of individual members. Balancing workload is expected to optimise levels of work stress throughout the construction development process. If appropriate, senior managers can also establish a stable working environment and progressively increase the autonomy of employees in handling projects. Hopefully, this will also enhance their sense of belonging (Peck 1998; Winter-Collins and McDaniel 2000).

Lastly, since prevention is always more important than cure, *regular assessments* of the work, emotional and physical stress experienced by construction personnel should be carried out. The results will not only help in assessing the well-being of individual employees and identifying those who need support, but also and perhaps more importantly, can highlight what kinds of stress management interventions, such as coaching or workshops on emotional and physical well-being, need to be offered to particular groups of staff.

References

Anderson, C.R. (1976) Coping behaviors as intervening mechanisms in the inverted-U stress performance relationship. *Journal of Applied Psychology*, 61(1), 30–34.

Anderson, S.M., Leu, J.R. and Kant, G.J. (1988) Chronic stress increases the binding of the A1 adenosine receptor agonist, [3H] cyclohexyladenosine, to rat hypothalamus. *Pharmacology Biochemistry and Behavior*, 30(1), 169–175.

Applebaum, H.A. (1999) *Construction Workers*. Westport, CN: Greenwood.

Barnes, G. (2009) Guess who's coming to work: Generation Y Are you ready for them?. *Public Library Quarterly*, 28(1), 58–63.

Baumeister, R.F. and Leary, M. R. (1995) The need to belong: Desire for interpersonal attachments as a fundamental human motivation. *Psychological Bulletin*, 117(3), 497–529.

Beehr, T.A., Walsh, J.T. and Taber, T.D. (1976) Relationship of stress to individually and organizationally valued states: Higher order needs as a moderator. *Journal of Applied Psychology*, 61(7), 41–47.

Beehr, T.A., Jex, S.M., Stacy, B.A. and Murray, M.A. (2000) Work stressors and coworker support as predictors of individual strain and job performance. *Journal of Organizational Behavior*, 21(4), 391–405.

Bishop, S.R., Lau, M., Shapiro, S., Carlson, L., Anderson, N.D., Carmody, J., Segal, Z.V., Abbey, S., Velting, D. and Devins, G. (2004) Mindfulness: A proposed operational definition. *Clinical Psychology: Science and Practice*, 11(3), 230–241.

Black, J.S. and Gregersen, H.B. (1990) Expectations, satisfaction and intention to leave of American expatriate managers in Japan. *International Journal of Intercultural Relations*, 14(4), 485–506.

Brunetto, Y. and Farr-Wharton, R. (2002) Using social identity theory to explain the job satisfaction of public sector employees. *International Journal of Public Sector Management*, 15(6–7), 534–551.

Blix, A.G., Cruise, R.J., Mitchell, B.M. and Blix, G.G. (1994) Occupational stress among university teachers. *Educational Research*, 36(2), 157–169.

Bowen, P., Edwards, P. and Lingard, H. (2013) Workplace stress experienced by construction professionals in South Africa. *Journal of Construction Engineering and Management*, 139(4), 393–403.

Cameron, N.M., Champagne, F.A., Parent, C., Fish, E.W., Ozaki-Kuroda, K. and Meaney, M.J. (2005) The programming of individual differences in defensive responses and reproductive strategies in the rat through variations in maternal care. *Neuroscience and Biobehavioral Reviews*, 29(4), 843–865.

Chartered Institute of Personnel and Development (2013) *Employee Turnover and Retention*, London: Chartered Institute of Personnel and Development.

Cherrington, D.J. (1994) *Organizational Behavior: The Management of Individual and Organizational Performance*. Needham Heights, MA: Paramount Publishing.

Choudhry, R.M. and Fang, D. (2008) Why operatives engage in unsafe work behavior: Investigating factors on construction sites. *Safety Science*, 46(4), 566–584.

Cockshaw, W.D. and Shochet, I. (2010) The link between belongingness and depressive symptoms: An exploration in the workplace interpersonal context. *Australian Psychologist*, 45(4), 283–289.

Cohen, S. (1980) Aftereffects of stress on human performance and social behavior: A review of research and theory. *Psychological Bulletin*, 88(1), 82–108.

Coomber, B. and Barriball, K.L. (2007) Impact of job satisfaction components on intent to leave and turnover for hospital-based nurses: A review of the research literature. *International Journal of Nursing Studies*, 44(2), 297–314.

Cooper, C.L. and Starbuck, W.H. (2005) *Work and Workers: Three-Volume Set*. London: Sage.

Cox, T. (1993) *Stress Research and Stress Management: Putting Theory to Work*. Sudbury: HSE Books.

Cropanzano, R., Rupp, D.E. and Byrne, Z.S. (2003) The relationship of emotional exhaustion to work attitudes, job performance and organizational citizenship behaviors. *Journal of Applied Psychology*, 88(1), 160–169.

Dainty, A.R.J., Bagilhole, B.M. and Neale, R.H. (2000) A grounded theory of women's career underachievement in large UK construction companies. *Construction Management and Economics*, 18(2), 239–250.

DeFrank, R.S. and Cooper, C.L. (1987) Worksite stress management interventions: Their Effectiveness and conceptualisation. *Journal of Managerial Psychology*, 2(1), 4–10.

Djebarni, R. (1996) The impact of stress in site management effectiveness. *Construction Management and Economics*, 14(4), 281–293.

Ducharme, L.J., Knudsen, H.K. and Roman, P.M. (2007) Emotional exhaustion and turnover intention in human service occupations: The protective role of coworker support. *Sociological Spectrum*, 28(1), 81–104.

Duffy, E. (1962) *Activation and Behavior*. New York: Wiley.

Feldman, D.C. (1976) A contingency theory of socialization. *Administrative Science Quarterly*, 21(3), 433–452.

Forman, S.G. (1981) Stress management training: Evaluation of effects on school psychological services. *Journal of School Psychology*, 19(3), 233–241.

Friend, K.E. (1982) Stress and performance: Effects of subjective work load and time urgency. *Personnel Psychology*, 35(3), 623–633.

Galegher, J., Kraut, R. and Egido, C. (1990) *Intellectual Teamwork: Social and Technological Foundations of Technical Work*. New York: Erlbaum.

Gmelch, W.H. (1982) *Beyond Stress to Effective Management*. New York: Wiley.

Gray-Toft, P. and Anderson, J.G. (1981) The nursing stress scale: Development of an instrument. *Journal of Behavioral Assessment*, 3(1), 11–23.

Hatton, C., Brown, R., Caine, A. andEmerson, E. (1995) Stressors, coping strategies and stress-related outcomes among direct care staff in staffed houses for people with learning disabilities. *Mental Handicap Research*, 8(4), 252–271.

Haynes, N.S. and Love, P.E. (2004) Psychological adjustment and coping among construction project managers. *Construction Management and Economics*, 22(2), 129–140.

Henne, D. and Locke, E.A. (1985) Job dissatisfaction: What are the consequences? *International Journal of Psychology*, 20(2), 221–240.

Holt, R.R. (1993) Occupational stress. In L. Goldberger and S. Breznitz (eds), *Handbook of Stress: Theoretical and Clinical Aspects*. New York: Free Press, 642–367.

Hong Kong Institute of Engineers (2013) *Graduate Scheme 'A' Training*. Retrieved from http://www.hkie.org.hk/eng/html/gradschemea/introduction.asp

Iaffaldano, M.T. and Muchinsky, P.M. (1985) Job satisfaction and job performance: A meta-analysis. *Psychological Bulletin*, 97(2), 251–273.

Jain, D. and Kobti, Z. (2011) Simulating the effect of emotional stress on task performance using OCC. In *Advances in Artificial Intelligence*. Heidelberg: Springer, 204–209.

Jamal, M. (1984) Job stress and job performance controversy: An empirical assessment. *Organizational Behavior and Human Performance*, 33(1), 1–21.

Jamal, M. (1999) Job stress and employee well-being: A cross-cultural empirical study. *Stress and Health*, 15(3), 153–158.

Jamal, M. (2007) Job stress and job performance controversy revisited: An empirical examination in two countries. *International Journal of Stress Management*, 14(2), 175.

Keenan, A. and Newton, T.J. (1987) Work difficulties and stress in young professional engineers. *Journal of Occupational Psychology*, 60(2), 133–145.

Krueger, G.P. (1989) Sustained work, fatigue, sleep loss and performance: A review of the issues. *Work and Stress*, 3(2) 129–141.

Lam, T., Lo, A. and Chan, J. (2002) New employees' turnover intentions and organizational commitment in the Hong Kong hotel industry. *Journal of Hospitality and Tourism Research*, 26(3), 217–234.

Lee, K.E. and Shin, K.H. (2005) Job burnout, engagement and turnover intention of dietitians and chefs at a contract foodservice management company. *Journal of Community Nutrition*, 7(2), 100.

Leung, M.Y. (2004) An international study on the stress of estimators. *The Hong Kong Surveyor*, 15(1), 49–52.

Leung, M.Y., Chan, Y.S., Chong, A. and Sham, J.F.C. (2008) Developing structural integrated stressor–stress models for clients' and contractors' cost engineers. *Journal of Construction Engineering and Management*, 134(8), 635–643.

Leung, M.Y., Chan, Y.S.I. and Dongyu, C. (2011) Structural linear relationships between job stress, burnout, physiological stress and performance of construction project managers. *Engineering, Construction and Architectural Management*, 18(3), 312–328.

Leung, M.Y., Chan, Y.S. and Olomolaiye, P. (2008) Impact of stress on the performance of construction project managers. *Journal of Construction Engineering and Management*, 134(8), 644–652.

Leung, M.Y., Chan, I.Y.S. and Yu, J.Y. (2009) Integrated model for the stressors and stresses of construction project managers. *Journal of Construction Engineering and Management*, 135(2), 126–134.

Leung, M.Y., Chan, I.Y.S. and Yu, J.Y. (2012) Preventing construction worker injury incidents through the management of personal stress and organizational stressors. *Journal of Accident Analysis and Prevention*, 48, 156–166.

Leung, M.Y., Chan, Y.S. and Yuen, K.W. (2010) Impacts of stressors and stress on the injury incidents of construction workers in Hong Kong. *Journal of Construction Engineering and Management*, 136(10), 1093–1103.

Leung, M.Y., Chen, D. and Yu, J. (2008) Demystifying moderate variables of the interrelationships among affective commitment, job performance and job satisfaction of construction professionals. *Journal of Construction Engineering and Management*, 134(12), 963–971.

Leung, M.Y., Liu, A.M.M. and Wong, M.K. (2006) Impacts of stress-coping behaviors on estimation performance. *Construction Management and Economics*, 24(1), 55–67.

Leung M.Y., Ng S.T. and Cheung S.O. (2002) Improving satisfaction through conflict stimulation and resolution in value management. *Journal of Management in Engineering*, 18(2), 68–75.

Leung M.Y., Ng S.T. and Cheung S.O. (2004) Measuring construction project participant satisfaction. *Construction Management and Economics*, 22(3), 319–331.

Leung, M.Y., Olomolaiye, P., Chong, A. and Lam, C.C. (2005) Impacts of stress on estimation performance in Hong Kong. *Construction Management and Economics*, 23(9), 891–903.

Ling, F.Y., Ibbs, C.W. and Hoo, W.Y. (2006) Determinants of international architectural, engineering and construction firms' project success in China. *Journal of Construction Engineering and Management*, 132(2), 206–214.

Locke, E.A. (1976) The nature and causes of job satisfaction. In M.D. Dunnette and L.M. Hough (eds), *Handbook of Industrial and Organizational Psychology*. Palo Alto, CA: Consulting Psychologists Press.

Lombardi, D., Verma, S., Brennan, M. and Perry, B. (2009) Factors influencing worker use of personal protective eyewear. *Accidents Analysis and Prevention*, 41(4), 755–762.

Loosemore, M. and Waters, T. (2004) Gender differences in occupational stress among professionals in the construction industry. *Journal of Management in Engineering*, 20(3), 126–132.

Louis, M.R. (1980) Surprise and sense making: What newcomers experience in entering unfamiliar organizational settings. *Administrative Science Quarterly*, 25(2), 226–251.

Maslow, A.H. (1943) A theory of human motivation. *Psychological Review*, 50(4), 370–396.

Mathieu, J.E. and Zajac, D.M. (1990) A review and meta-analysis of the antecedents, correlates and consequences of organizational commitment. *Psychological Bulletin*, 108(2), 171–194.

Maurya, S.S. (2012) Job satisfaction of managers, engineers: A case study in the UAE. *International Journal of Business Economics and Management Research*, 2(8), 134–144.

McGrath, J.E. (1976) Stress and behavior in organizations. In M.D. Dunnette (ed.), *Handbook of Industrial and Organizational Psychology*. Chicago, IL: Rand McNally.

Mccoll, M.A., Lei, H. and Skinner, H. (1995) Structural relationships between social support and coping. *Social Science and Medicine*, 41(3), 395–407.

Mearns, K., Flin, R., Gordon, R. and Fleming, M. (2001) Human and organizational factors in offshore safety. *Work and Stress*, 15(2), 144–160.

Meglino, B.M. (1977) The stress–performance controversy. *MSU Business Topics*, 25, 53–59.

Milliken, T.F., Clements, P.T. and Tillman, H.J. (2007) The impact of stress management on nurse productivity and retention. *Nursing Economics*, 25(4), 203–210.

Moorman, R.H. (1993) The influence of cognitive and affective based job satisfaction measures on the relationship between satisfaction and organizational citizenship behavior. *Human Relations*, 6, 759–776.

Mullen, J. (2004) Investigating factors that influence individual safety behaviour at work. *Journal of Safety Research*, 35(3), 275–285.

Murray, M. and Zagaretos, P. (2001) The influence of cultural differences on the performance of international contractors. In *Proceedings of the Seventeenth Annual Conference of ARCOM, Association of Researchers in Construction Management*, 101–110.

Muse, L.A., Harris, S.G. and Feild, H.S. (2003) Has the inverted-U theory of stress and job performance had a fair test? *Human Performance*, 16(4), 349–364.

Nandram, S.S. and Klandermans, B. (1993) Stress experienced by active members of trade unions. *Journal of Organizational Behavior*, 14(5), 415–431.

O'Reilly, C.A., Chatman, J. and Caldwell, D.F. (1991) People and organizational culture: A profile comparison approach to assessing person-organization fit. *Academy of Management Journal*, 34(3), 487–516.

Peck, S.R. (1998) No easy roads to employee involvement. *Academy of Management Executive*, 12(3), 83–84.

Pflanz, S.E. and Ogle, A.D. (2006) Job stress, depression, work performance, and perceptions of supervisors in military personnel. *Military Medicine*, 171(9), 861–865.

Quick, J.C., Quick, J.D., Nelson, D.L. and Hurrell, J J. (1997) *Preventive Stress Management in Organizations*. Washington, DC: American Psychological Association.

Reichers, A.E., Wanous, J.P. and Steele, K. (1994) Design and implementation issues in socializing (and resocializing) employees. *Human Resource Planning*, 17, 17–25.

Robbins, T.W. and Fray, P.J. (1980) Stress-induced eating: Fact, fiction or misunderstanding? *Appetite*, 1(2), 103–133.

Robbins, S.P. and Judge T.A. (2003) *Organizational Behaviors*. Toronto: Pearson Prentice Hall.

Robertson, I. and Cooper, C.L. (2012) *Wellbeing: Productivity and Happiness at Work*. Basingstoke: Palgrave Macmillan.

Schuler, R.S. (1982) An integrative transactional process model of stress in organizations. *Journal of Organizational Behavior*, 3(1), 5–19.

Scott, L.D., Rogers, A.E., Hwang, W.T. and Zhang, Y. (2006) Effects of critical care nurses' work hours on vigilance and patients' safety. *American Journal of Critical Care*, 15(1), 30–37.

Selye. H. (1974) *Stress without Distress*. Philadelphia: Lippincott.

Selye, H. (1976) *Stress in Health and Disease*. Boston: Butterworths.

Selye, H. (1982) History and present status of the stress concept. In L. Goldberger and S. Breznitz (eds), *Handbook of Stress: Theoretical and Clinical Aspects*. New York: Free Press, 7–17.

Semmer, N. (2003) Individual differences, work stress and health. *Handbook of Work and Health Psychology*, 2, 83–120.

Shore, L.M. and Martin, H.J. (1989) Job satisfaction and organizational commitment in relation to work performance and turnover intentions. *Human Relations*, 42(7), 625–638.

Spatz, D.M. (2000) Team-building in construction. *Practice Periodical on Structural Design and Construction*, 5(3), 93–105.

Tarafdar, M., Tu, Q., Ragu-Nathan, B.S. and Ragu-Nathan, T.S. (2007) The impact of technostress on role stress and productivity. *Journal of Management Information Systems*, 24(1), 301–328.

Trzcieniecka-Green, A., Gaczek, A., Pawlak, A., Orlowska, W. and Pochopin, T. (2012) The sense of life satisfaction and the level of perceived stress in the midwifery profession – A preliminary report. *Achieves of Psychiatry and Psychotherapy*, 3, 35–43.

US Department of Labor. (2007) *Working Conditions in Construction Industry*. Retrieved from http://www.bls.gov/oco/cg/cgs003.htm

Vroom, V.H. (1982) *Work and Motivation*. Malabar, FL: Robert E. Krieger.

Westman, M. and Eden, D. (1991) Implicit stress theory: The spurious effects of stress on performance ratings. *Journal of Social Behavior and Personality*, 6(7), 127–140.

Winter-Collins, A. and McDaniel, A.M. (2000) Sense of belonging and new graduate job satisfaction. *Journal for Nurses in Staff Development*, 16(3), 103–111.

Wright, T.A. and Cropanzano, R. (1998) Emotional exhaustion as a predictor of job performance and voluntary turnover. *Journal of Applied Psychology*, 83, 486–493.

Yang, N.Y. and Moon, S.Y. (2011) Relationship of self-leadership, stress and satisfaction in clinical practice of nursing students, *Journal of Korean Academy of Nursing Administration*, 17(2), 216–225.

Yang, Y.F. (2009) An investigation of group interaction functioning stimulated by transformational leadership on employee intrinsic and extrinsic job satisfaction: An extension of the resource-based theory perspective. *Social Behavior and Personality*, 37(9), 1259–1277.

Yerkes, R.M. and Dodson, J.D. (1908) The relation of strength of stimulus to rapidity of habit-formation. *Journal of Comparative Neurology and Psychology*, 18(5), 459–482.

Zaccaro, S.J. and Riley, A.W. (1987) Stress, coping and organizational effectiveness. *Occupational Stress and Organizational Effectiveness*, 1–28.

Zajonc, R.B. (1965) *Social Facilitation*. Research Center for Group Dynamics, Institute for Social Research, University of Michigan.

6

Stress Management

6.1 Coping Behaviours

When talking about 'stress', people generally mean overstress and its effects. Although too much stress (overstress) can result in burnout, too little stress (understress) can also affect the performance of construction professionals through rustout (Haynes and Love 2004; Lingard 2003). Therefore, stress often plays a harmful role in terms of performance and, simultaneously, a motivating role for improving performance. Construction personnel's stress should not be simply eliminated completely, but managed in order to optimise their performance.

Extending this to some circumstances means balancing various aspects of individual life, including work, relationships and leisure, as well as balancing the physical, intellectual and emotional aspects of life. For instance, people who are successful in managing stress think of life as a challenge rather than as a series of irritations. Human beings react to stress with different behaviours (Cohen, Kessler and Gordon 1995; Keavney and Sinclair 1978). Reactions to stress are not simply 'what you should, would or could do', but 'what you in fact do as you react to particular conditions and demands' (Schafer 2000). Thus, stress management is an ongoing and dynamic process, involving the appraisal process for the stress and one's own possibilities for dealing with it and the strategies to handle it (Lösel and Bliesener 1990). Hence, 'coping with stress' refers to people's cognitive and behavioural efforts to manage the environment, internal demands and the conflicts between them (Djebarni 1996; Lazarus and Folkman 1984).

Among the various coping theories, the transactional model of stress and coping developed by Lazarus and Folkman (1984) is one of the most widely

Stress Management in the Construction Industry, First Edition.
Mei-yung Leung, Isabelle Yee Shan Chan and Cary L. Cooper.
© 2015 John Wiley & Sons, Ltd. Published 2015 by John Wiley & Sons, Ltd.

recognised and used. According to this model, there are three stages of coping: primary appraisal, secondary appraisal and coping. In the primary appraisal stage, people decide whether they are potentially threatened or in jeopardy in a particular situation that is worth being concerned about. The coping process will end if the situation is judged to be irrelevant or trivial. In contrast, if the circumstance is meaningful and potentially threatening, the stress-coping process will continue into the secondary appraisal stage, in which people assess their resources for dealing with the stressors. This stage depends on the previous experiences of the individual in similar environments, generalised beliefs about oneself and the environment, the availability of resources to deal with the situation and an assessment of the degree of control over the situation (Holroyd and Lazarus 1982; Schafer 2000). The lesser the available resources and the perceived control, the more threatening the situation is and thus the greater the probability of mental and physical distress. When people perceive a situation to be taxing or exceeding their ability to adapt, the process continues to coping, the third stage, in which they take whichever actions seem appropriate to deal with the situation (Stein et al. 1997). This can include physical actions and cognitive adjustments. However, the effectiveness of these actions is another matter (Frese 1986; Krohne 1986; Laux 1986; Schafer 2000).

Based on sociological and psychological aspects, there are three main types of coping behaviours among construction personnel: problem-based coping, emotion-based coping and meaning-based coping (Folkman 2010; Lazarus and Folkman 1984). Lazarus and Folkman first distinguished problem-based coping from emotion-based coping. Problem-based coping is doing something to specifically deal with the stressful problem or situation. Emotion-based coping is when people do something to make themselves feel better about the situation. Later on, Folkman (1997) added meaning-based coping to cover coping that results in positive emotion and re-enacts the adoption of problem-based or emotion-based coping – the key to sustaining a coping process.

6.1.1   Problem-Based Coping Behaviours

Problem-based coping behaviours involve efforts to deal with the sources of stress through the modification of their own problem-maintaining behaviours or the environment (Djebarni 1996; Greenglass, Schwarzer and Taubert 1999; Leung, Wong and Oloke 2003; Leung, Liu and Wong 2006). These efforts focus on dealing constructively with the stressors or the circumstance itself (Folkman, Schaefer and Lazarus 1979; Pearlin and Schooler 1978; Schafer 2000). This type of coping effort may be directed at the situational demands and problem solving (Thoits 1995). One study found that construction project managers who engaged in more problem-based coping, such as active coping, were better adjusted to stressful, changing work environments than managers who engaged in more emotion-based coping (Haynes and Love 2004). The next sections describe various types of problem-based coping behaviours.

6.1.1.1 *Planful Problem Solving*

Planful problem solving is an essential aspect of problem-based coping. It is self-initiated, overt behaviour that directly deals with the problem and its effects (Lazarus and Folkman 1984). In general, people use planful problem solving when they believe they can influence the environment. This strategy reflects active approaches to making a plan of action, focusing efforts to solve the problem at hand and taking direct action (Newton and Keenan 1985). It is associated with decreased stress levels and improved health (Israel, Schurman and House 1989; Spector 1986). Planful problem solving was found to be the most frequently used coping method among construction personnel (Yip and Rowlinson 2006). In fact, a study found that taking direct action towards stressful problems is beneficial to the performance of quantity surveyors and project managers in construction projects and in their personal lives (Leung, Liu and Wong 2006). Another study found that the primary strategy construction professionals used in planful problem solving was to deal with problem directly (Chan, Leung and Yu 2012).

6.1.1.2 *Positive Reappraisal*

Positive reappraisal simply refers to restructuring a problem positively, which has been found to be associated with decreased depressive symptoms (Holahan et al. 1997; Greenglass, Schwarzer and Taubert 1999). In fact, being objective about problems is an effective way for project managers to cope with stress (Rivera 2008). Positive reappraisal is an internal coping strategy in which an individual implements intrapsychic and cognitive adjustments to the situation (Lazarus and Folkman 1984). It is a frequently used strategy and the only coping strategy that remains stable over time (King 1985; Stewart et al. 1997). It has also been found that situational appraisals of control are linked to performance of active problem-solving coping strategies (Folkman, Aldwin and Lazarus 1981). In addition, a study found that construction project managers use positive reappraisal and that it is a predictor of their adjustment (Haynes and Love 2004).

6.1.1.3 *Confrontive Coping*

Confrontive coping refers to aggressive efforts to alter the situation, including a certain degree of hostility and risk taking (Folkman et al. 1986). It deals with self-regulatory goal attainment processes and explains what motivates people to strive for ambitious goals and to commit themselves to improve their performance. Therefore, confrontive coping is an autonomous and self-determined goal-setting process (Greenglass, Schwarzer and Taubert 1999; Lazarus and Folkman 1984; Schwarzer 1999). For instance, construction personnel who encounter inadequate resources in a construction project and use this coping strategy may try to get their superiors or departments with resources (i.e., the people responsible for the problem) to change their course of action or change their mind in using alterative construction methods or rearranging manpower. Previous studies have confirmed that the confrontive coping strategy is effective in dealing with stress at work (Chung et al. 2010; Xianyu and Lambert 2006). Therefore, researchers have suggested that construction professionals in Ireland and Hong Kong should use confrontive coping strategies to improve their performance (Chan, Leung and Yu 2012; Gunning and Keaveney 1998).

6.1.1.4 *Instrumental Support Seeking*

Seeking instrumental support simply refers to people looking for instrumental advice, assistance or information from their social network, including colleagues, people with more experience, or friends to solve the problem (Billings and Moos 1984; Carver, Scheier and Weintraub 1989; Leung, Wong and Oloke 2003; Leung, Liu and Wong 2006; Varhol 2000). It is the most frequently used strategy among people suffering from stress and it is found effective at relieving stress (Holahan et al. 1997; Stewart et al. 1997). Through instrumental support seeking, people can discuss their problems and get advice that is essential in solving and handling their stress. This strategy is used among the general public (Wills 1991), teachers (Greenglass 1997), quantity surveyors (Leung, Liu and Wong 2006) and other construction professionals (Chan 2011).

6.1.1.5 *Active Coping*

Active coping is the process of taking active steps to remove or circumvent stressors or to ameliorate the effects (Carver, Scheier and Weintraub 1989). In general, this control-directed coping effort will only be effective when people perceive that they actually have the ability to control the stressors, either on a cognitive or behavioural level (Latack 1986). One study found that active coping moderates the interaction between job demands and job control (de Rijk et al. 1998) and is necessary for construction project managers to relieve their stress (Haynes and Love 2004). However, it is also worth noting that an active strategy is likely to be counterproductive if people do not believe they have control over the situation (Lennerlöf 1988).

6.1.2   **Emotion-Based Coping Behaviours**

Emotion-based coping behaviour involves coping efforts that are aimed at adjusting emotional distress and maintaining a moderate level of motivation (Djebarni 1996; Greenglass, Schwarzer and Taubert 1999; Leung, Wong and Oloke 2003; Leung, Liu and Wong 2006). It focuses on dealing with an individual's own fears, anger or guilt as one reacts to a situation (Folkman,

Schaefer and Lazarus 1979; Pearlin and Schooler 1978; Schafer 2000). It aims at managing distressing emotions through various kinds of behaviour (Djebarni 1996; Latack and Havlovic 1992).

6.1.2.1 *Emotional Support Seeking*

Seeking emotional support refers to getting moral support, sympathy, comfort or understanding from family, friends or even religion (Carver, Scheier and Weintraub; Newton and Keenan 1985). Actually, it is similar to one of the coping strategies under the category of problem-based coping – 'seeking social support'. However, this coping behaviour is mainly concerned with people's subjective feelings and emotional release, rather than instrumental support, which aims at dealing with the situation effectively. Previous studies found that this coping behaviour alleviates the stress of quantity surveyors, particularly junior quantity surveyors (Leung, Liu and Wong 2006) and can have significant effects on the emotional exhaustion of employees in the construction industry (Francis and Lingard 2004). Furthermore, there is evidence that women civil engineers use this strategy to balance work–family demands (Watts 2007).

6.1.2.2 *Emotional Discharge*

Emotional discharge is a behavioural expression of unpleasant emotions to reduce tension (Billings and Moos 1984; Leung, Wong, and Oloke 2003; Leung, Liu and Wong 2006). Manifestations of this coping skill include exercising, eating, smoking, drinking and taking drugs (Billings and Moos 1984; Schafer 2000). In fact, the coping behaviours of emotional discharge such as smoking and drinking significantly increase when an individual encounters stressful situations, such as adjusting to a new job and lifestyle after expatriation (Anderzen and Arnetz 1997). In a recent study in construction, alcohol consumption was found to be widespread, with more than one-third of respondents consuming 3–9 units/week, while one in six respondents was found to smoke up to 40 cigarettes per day (Bowen et al. 2014). However, this coping behaviour was found to be detrimental to individual performance, although it can alleviate the negative impact of work stress on the interpersonal performance of construction professionals (Chan 2008; Leung, Liu and Wong 2006).

6.1.2.3 *Escapism/Denial*

Escapism coping has been conceptualised as containing elements of people-oriented, situation-oriented and task-oriented avoidance responses (Leung, Wong and Oloke 2003; Parker and Endler 1996). In other words, escapism coping refers to an individual escaping from the people, situation or tasks that induce stress in them. This coping behaviour leads to increased levels of depression among older individuals and increased severity of psychopathology, symptoms of depression and emotional distress (Greenglass, Schwarzer and Taubert 1999). Denial coping refers to isolating oneself from the stress by ignoring the problem (Carver, Scheier and Weintraub 1989; Newton and Keenan 1985). There are several types of denial, such as denial of illness, denial of impact, denial of affect and suppression of thoughts about the stressful situation (Havik and Maeland 1988; Lowery et al. 1992). A study found that higher levels of denial among individuals are negatively related to anxiety and depression, as well as general measures of psychopathology (Esteve et al. 1992).

6.1.2.4 Self-Controlling

Self-controlling refers to managing the stress reaction by suppressing emotions and actions (Lazarus and Folkman 1984). When facing a stressful situation, people attempt not to feel or show emotional reactions; they try to stay calm, not become frustrated; and they take no concrete action to tackle the real problem. Researchers have recommended this coping behaviour for construction professionals who deal with stress issues (Gunning and Keaveney 1998). Although there is a study claiming that people who use self-controlling coping have less stress because they are able to detach themselves from the situation or create a positive outlook (Laranjeira 2012), another study found that people who use self-control tend to have poorer mental health outcomes (Chang et al. 2006).

6.1.2.5 Accepting Responsibility

Accepting responsibility is a coping method in which people acknowledge their own obligations and accept their own duties in the stressful event (Lazarus and Folkman 1984). When people use this coping method, they accept the fact that the stressful event has occurred and is real. They try to acknowledge their contribution to the problem with a concomitant theme of trying to overcome the situation, such as apologies or taking a remedial action. With a sense of blaming oneself but without modifying the situation itself, this coping behaviour has been found to be associated with increased stress (Folkman and Lazarus 1988), increased depression (Madu and Roos 2006) and reduced subjective well-being (Glidden, Billings and Jobe 2006).

6.1.3   Meaning-Based Coping Behaviours

Meaning-based coping focuses on the positive emotions, which sustains problem-focused and emotion-focused coping behaviours. Folkman (1997) identified various meaning-based coping strategies, such as spiritual beliefs and practices and the infusion of ordinary events with positive meaning. These strategies are closely related to people's psychological mind-set and the unity of one's body, which coincide with the core of the mindfulness-based stress reduction (MBSR) strategies developed by Kabat-Zinn (1982).

To restore a balanced sense of health and well-being, people need increased awareness of all aspects of the self, including body, mind and heart. The goal of MBSR is to ignite this inner capacity and infuse people's life with awareness (Kabat-Zinn et al. 1992). Mindfulness is a state of consciousness that involves consciously attending to one's moment-to-moment experience (Brown and Ryan 2003). It is not something that we have to 'get' or acquire; it is already within us – a deep internal resource available and patiently waiting to be released and used in the service of learning, growing and healing. MBSR consists of a systematic and progressive programme that aims to alleviate individual physical and psychological disorders through mindfulness skills in a sustainable way (Grossman et al. 2004). MBSR focuses on the progressive obtainment of mindful and non-evaluative awareness in various aspects of individual life, including physical sensations, perceptions, affective states, thoughts and imagery. Through observing one's mental and bodily responses to external stimuli in every present moment, people have a sense of control over their body, mind and heart through the cultivated clarity, insights and understanding of individual life.

6.1.3.1 *MBSR Attitudes/Behaviours*

In general, there are 8 attitudes and behaviours that are cultivated in association with the core attitude of MBSR (see also the next section for MBSR practices). These attitudes include:

1. *Present focus.* Present focus is one of the core factors of mindfulness. It refers to a sense of direct and immediate experience of the present moment without becoming distracted by or lost in thoughts and feelings and without identifying with those mental states.
2. *Beginner's mind.* Rather than observing experience through the filter of our beliefs, assumptions, expectations and desires, mindfulness involves a direct observation of various objects as if for the first time, a quality that is often referred to as 'beginner's mind'.
3. *Non-judgmental.* It is easy to discover that part of our mind is constantly evaluating our experience, comparing it with other experiences or holding them up against expectations and standards that we create, often out of fear. Being non-judgmental simply means trying not to evaluate things with labels, such as good or bad, right or wrong, worthwhile or worthless.
4. *Acceptance.* Acceptance is being experientially open to the reality of the present moment, which involves a conscious decision to abandon one's fixed mind-set in order to have a different experience and an active process of 'allowing' current thoughts, feelings and sensations. With acceptance, people should also be patient towards the present moment and not have to let their anxieties and desire for certain results dominate the quality of the moment, even when things are painful.

5. *Describing.* Describing refers to being able to use words to describe, note or label a stimuli or phenomenon, such as emotions, sights, cognitions, smells and sound.

6. *Awareness.* To act with awareness is to wholly participate in a current activity with undivided attention as opposed to behaving automatically or absent-mindedly. Mindful awareness is about being careful with one thing at a time – either internal or external phenomena, such as thoughts, ideas, images, emotions, sensations and movements.

7. *Non-reactivity.* In a state of mindfulness, thoughts and feelings are observed as events in the mind, without over-identifying with them and without reacting to them in an automatic, habitual pattern of reactivity. This dispassionate state of self-observation is thought to introduce a 'space' between one's perception and response.

8. *Letting go.* Letting go is when people cease clinging to things, such as ideas, thoughts, feelings and desires. It is a key skill in avoiding vicious cycles.

6.1.3.2 MBSR Practices

MBSR emphasises the experiential cultivation of both *formal* and *informal* mindfulness practices, such as mindfulness body scanning and meditation. These practices act as a foundation for an individual to cultivate positive health behaviours and psychological and emotional resilience that can be used effectively across the lifespan. It is a way to facilitate insight into the impermanent nature of the personal self via awareness. Insight is posited to occur through the recognition of conditioned chains of mental processes and the attendant woes that follow from these (Carmody et al. 2009). Mindfulness can be achieved through meditation (i.e., formal practices), but people can also practise mindfulness through daily living (i.e., informal practices). Daily exercises help us to implement MBSR practices in daily life to handle problems by ourselves (Hawley et al. 2014).

6.1.3.2.1 Formal MBSR Practices

In order to increase their awareness, construction personnel can try to feel the sensations of their body and breathing including static states and movement states. Formal MBSR practices consist of body scanning, breath, meditation, sitting meditation, walking meditation and mindful yoga (Kabat-Zinn 2003).

1. *Body scanning.* Body scanning can be used to relieve physical pain (Ott, Longobucco-Hynes and Hynes 2002). Paradoxically, focused awareness on specific pain as it occurs in the body allows people to experience the subtle shifts and changes that happen during the subjective experience of pain in that moment. As a result, people avoid magnifying pain by not getting caught up in stories and memories of previous pain or anticipating prolonged or future pain. Instead, attention is focused on what is happening in the present moment with any physical sensation.

People begin to learn the subtle cues from the body that indicate an imbalance or the need for a particular intervention, such as a change in position or rest (Olpin and Hesson 2012).

2. *Breath meditation.* Breath is life; as breathing fails, life slowly ebbs away (McDonald and Stocks 1965). Being aware of our breath is one of the simplest ways to calm down and integrate the mind and body. Most of us are unconscious of our moment-to-moment breathing because breathing is the most basic and essential function of the body and hence of the story of our life as it unfolds (Holmes 2007). By consciously observing our inhalations and exhalations, we tune our awareness into the vital realm of consciousness that is alive in the here and now. With persistence and consistency of practice, breath meditation makes people more aware of their bodily sensations. Meditation involves letting go of wandering thoughts or ideas and focusing on our breathing (Robins 2003).

3. *Sitting meditation.* In sitting meditation, we allow and accept whatever we observe and feel (i.e., acceptance). Practitioners first find a sitting posture that embodies a sense of calmness and dignity (Segal, Williams and Teasdale 2012). The focus of attention can be moved from breathing to the body, sound and thought. Whenever wandering thoughts or ideas come up, practitioners can invite and remind themselves to bring their attention back to their breath (Moore 2008). The emphasis is on simply taking notice of whatever the mind happens to wander to and accepting each object without judging it or elaborating on its implications, additional meanings or need for action (Kabat-Zinn 1990; Segal, Williams and Teasdale 2002).

4. *Walking meditation.* Walking meditation has been described as 'meditation in motion': being with each step, walking for its own sake, without any destination (Segal, Williams and Teasdale 2002). Mindful walking takes the everyday activity of walking and uses it as a mindfulness practice to become more aware of bodily sensations (Segal, Williams and Teasdale 2002). We walk, knowing that we are walking, feeling the walking. The physical sensation of walking enables people to feel more 'grounded' in the present moment, especially for construction personnel who often work in construction sites. When people's minds are full and they are stressed, paying attention to and being aware of their physical movements can be an easy way for them to be mindful.

5. *Mindful yoga.* Mindful yoga refers to safe, gentle and slow yoga stretching exercises that focus on moment-to-moment awareness to breath, physical movement, bodily sensation, thoughts and emotions that happen during the practice (Kabat-Zinn 2003; Salmon et al. 2009). When stretching the body, people focus their mind on how their breath changes to accompany the stretching of different muscles. Through this practice, construction personnel can integrate and be more aware of their body, mind and spirit, which improves health and vitality. In addition to promoting stress resilience, mindful yoga has also been found to relieve musculoskeletal problems, such as reducing stiffness and increasing flexibility (Quinn, Bussell, and Heller 2010).

6.1.3.2.2 Informal MBSR Practices

In addition to the formal MBSR practices above, the attitudes of MBSR can also be brought into and practised in one's daily activities, including mindful eating, three-minute breathing, communication and smiling (Kristeller and Hallett 1999).

1. *Mindful eating.* Mindful eating invites us to eat slowly, paying attention to the process of eating, which engages all parts of our body, heart and mind (Bays 2009). This process emphasises moment-to-moment awareness of the appearance, shape, smell, sound, taste and texture of the food and our bodily sensations from the beginning to the end of eating. The mindful eating process not only allows us to eat slowly, which is beneficial to our physical health, but also cultivates gratitude and appreciation for the invaluable food, helping developing one's positive psychology for stress resilience (Kristeller and Hallett 1999; Mathieu 2009).

2. *The three-minute breathing space.* The three-minute breathing space is a practical, effective tool for managing stress by feeling more centred in the present moment. This practice involves: (i) *acknowledging* (awareness) one's present experience; (ii) *gathering* attention to the speed, quality and sensation of breathing through normal inhalation and exhalation; and (iii) *expanding* the awareness of breathing to the whole body, including body posture, sensations, facial expressions and so on (Crane 2009; Germer, Siegel and Fulton 2005; Megginson and Clutterbuck, 2012). This practice can be a habit or tool to help people be anchored in the present moment and it can be grounded in their thoughts and emotions (Segal, Williams and Teasdale 2012).

3. *Mindful communication.* Because our thoughts easily wander and ideas arise in our minds, it is often hard for us to pay attention to and not interrupt others when they are talking to us. This influences not only the context of dialogue, but also the interpersonal connection and relationships between people. In general, communication is two way: sending messages (speaking, written and nonverbal acts) and receiving them (listening, reading and observing the behaviour of others). Mindful listening emphasises empathy, which helps people express both respect and compassion for others (Borker 2012). The practices of mindful listening remind practitioners to notice any resistance, tensing or rushing in the conversation (Heaversedge, 2012), bring attention to listening and avoid missing important information from others in conversations (Stahl and Goldstein 2010). In such situations, construction personnel can then remain present with the self and others, create space for a response that comes from the head rather than the body. Hence, mindful communication can lower individual reactivity to stressful events and help people be more resilient in the face of adversity (Krasner et al. 2009).

4. *Smiling.* Smiling is a universal phenomenon (Lefrancois 1996). Mindful smiling is a form of a moving meditation. It can inspire peace,

comfort and even strength. Finding a soft smile at times when energy is low and the body is challenged can help people conjure up strength. It reconnects practitioners with their determination and reminds them of individual physical, mental and societal opportunities to practise smiling freely. The practise of mindful smiling is deepened by bringing awareness to the breath. As you inhale, feel and think about the fact you are smiling. As you exhale, say to yourself, 'let go' or 'release'. The smile does not need to be big or toothy, but regularly practising mindful smiling can refresh our mind and build good relationships with others (Hanh 1998).

6.2   Effectiveness of Coping Behaviours

Coping behaviour is a straightforward and inexpensive way for people to manage their stress because it does not disrupt production schedules or create upheaval in the organisational structure (Murphy 1988). By providing psychological and physical benefits to individuals, coping behaviour is generally effective (Pearlin and Schooler 1978; Sallis et al. 1987). It also improves work efficiency, improves interpersonal relationships and, more importantly, reduces absenteeism, turnover and medical care costs of the organisation. Effective coping behaviour can maximise positive consequences, minimise the impact of stress and alleviate its negative consequences at the same time (Lazarus and Folkman 1984). However, not all coping behaviours are effective in relieving stress. Coping can be adaptive or maladaptive. Adaptive coping is when people deal effectively with stressful situations and minimise distress. This method can improve personal health, productivity, life satisfaction and personal growth. In contrast, maladaptive coping leads to unnecessary distress for the individual and erodes wellness (Schafer 2000). In view of this, previous studies have been conducted to investigate the effectiveness and impact of various coping behaviours.

6.2.1   Effectiveness of Problem-Based Coping

Problem-based coping behaviours have long been identified as effective coping strategies because they aim to solve the stressful problem directly, such as reducing uncertainty or ambiguity at work (Christman et al. 1988; Livneh 1999; Webster and Christman 1988). People who use problem-based coping behaviours are found to suffer from fewer harmful physical outcomes (e.g., headaches and physical fatigue) and psychological outcomes (e.g., emotional distress, job anxiety and depressive symptoms) than people who use emotion-based strategies, such as passive avoidance (Fincham, Beach and Davila 2004; Dyson and Renk 2006; Folkman and Moskowitz 2000).

6.2.2   Effectiveness of Emotion-Based Coping

Although some studies have found that emotional support seeking, as an emotion-based coping behaviour, alleviates stress and improves performance (Kraimer, Wayne and Jaworski 2001; Wang and Nayir 2006), emotion-based coping is generally considered an ineffective strategy because treating the emotion alone does not solve the problem. Hence, studies normally find that emotion-based coping behaviour is associated with increased depressive symptoms (Dyson and Renk 2006), increased psychological symptomatology, disruption of social and recreational activities, poor personal ratings of global coping effectiveness (Livneh 1999; Terry 2011), increased uncertainty or ambiguity at work (Christman et al. 1988; Livneh 1999; Webster and Christman 1988) and decreased job satisfaction (Greenglass, Schwarzer and Taubert 1999). Although there is no definite conclusion on whether emotion-based coping behaviours are effective or not, emotion-based coping is generally recognised to be an ineffective coping strategy.

6.2.3   Effectiveness of Meaning-Based Coping (MBSR)

Studies have conclusively demonstrated the positive effects of MBSR in relieving stress, mitigating depression, reducing psychological stress symptoms (e.g., anxiety), improving physical functioning (e.g., relieving chronic pain, reducing the frequency of medical visits and reducing eating disorders), improving task and interpersonal performance (Brown and Ryan 2003; Carmody and Baer 2008; Franco et al. 2010; Jain et al. 2007; Kristeller and Hallett 1999; Morone, Greco and Weiner 2008). However, research on MBSR is still very rare in the construction industry. In view of the effectiveness of MBSR in other sectors, it is worthwhile to bring this new stress management concept to the construction sector.

6.3   Determinants of Various Coping Behaviours

If there are adaptive and maladaptive coping styles, what determines which coping behaviours people use? There are different schools of thoughts on this question. First, some researchers have proposed that people with different work background and characteristics (such as their work experience and control over work) use different coping behaviours. Senior people normally use problem-based coping behaviour to alleviate the stress they encounter in their jobs, whereas junior people tend to use emotion-based coping methods to relieve stress (Leung, Liu and Wong 2006). People who have little control over their work are less likely to engage in problem-based coping and are more likely to use emotion-based strategies (Folkman 1984).

Second, some researchers focus on the influence of personal traits on coping behaviours. Individuals with high self-efficacy are more likely to feel that they are able to control challenging environmental demands by taking adaptive action (Bandura 1992). When people believe that outcomes are within their control, they are more likely to use problem-based coping behaviour than people who see outcomes as resulting by chance (Bandura 1992; Folkman 1984; Schafer 2000). Moreover, personal traits also affect the application of meaning-based coping behaviour. For instance, people with a proactive personality are more likely to cope with stress mindfully (de Ribaupierre 2000). On the other hand, the different emotional states of individual may also affect their adoptions of different coping behaviours (Bishop 2002). It is relatively harder for an irritated, anxious person to calm down, focus on the present moment and adopt the meaning-based coping behaviour to mitigate stress.

Third, studies have shown that culture and cultural values influence people's coping behaviours (Lam and Zane 2004; See and Essau 2010; Xianyu and Lambert 2006). For instance, construction professionals who place importance on the Chinese cultural values of interpersonal integration tend to use planful problem solving, while people who emphasise the discipline work ethos tend to use more positive reappraisal and less emotional discharge (Chan and Leung 2013). Moreover, Western and Eastern people take different approaches to dealing with stress: Westerners are more likely to use problem-based coping behaviours, such as planful problem solving and positive reappraisal (Hoedaya and Anshel 2003; Tweed, White and Lehman 2004), whereas Easterners tend to take more emotion-based approaches, such as escapism and acceptance of responsibility (Hoedaya and Anshel 2003; O'Connor and Shimizu 2002; Tweed, White and Lehman 2004). Studies that provide insight into the determinants of coping styles can help us guide people towards effective coping styles.

6.4   Developing a Conceptual Model of the Individual Coping Behaviours of Construction Personnel

Based on the review above, it can be summarised that there are three main types of individual-level coping behaviours that construction personnel can use: problem-based coping behaviours, emotion-based coping behaviours and meaning-based coping behaviours (MBSR). These three types of coping behaviours deal with stress in different ways: problem-based coping emphasises *problem* solving, emotion-based coping emphasises dealing with *emotional* states and MBSR emphasises cultivating the stress-resilience attitudes and behaviours of the *body* and *mind*. A comprehensive stress management approach should encompass all these different aspects. Figure 6.1 illustrates a model of individual coping behaviour, including three coping behaviours and their various sub-coping behaviours.

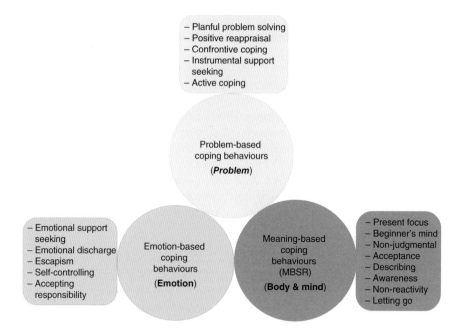

Figure 6.1   Conceptual model of individual coping behaviours for construction personnel.

6.5   Studies on the Coping Behaviours of Construction Personnel

Following the literature review of various coping behaviours, this section aims to further explore and provide scientific support for the experiences of various types of coping behaviours among construction personnel. To achieve this aim, this study used both statistical analyses and case studies.

6.5.1   Questionnaire

6.5.1.1   *Coping Behaviours Used by Construction Personnel*

To measure the coping behaviours of construction professionals, 64 items were adopted from our previous studies, covering different types of problem-based, emotion-based (Leung, Wong and Oloke 2003; Leung, Liu and Wong 2006) and meaning-based (MBSR) coping behaviours. Respondents were asked to rate their agreement with statements about the adoptions of various coping behaviours on a 7-point Likert scale from 1 (*strongly disagree*) to 7 (*strongly agree*). Based on a total of 702 participants, 15 types of coping behaviours were classified into the three groups (i.e., problem-based, emotion-based and meaning-based) and analysed.

Problem-based coping behaviours include four factors: planful problem solving (planful problem solving and active coping were combined into one factor), instrumental support seeking, confrontive coping and positive reappraisal. Emotion-based coping behaviours covered four factors: emotional support seeking, emotional discharge, escapism and self-controlling; in which accepting responsibility is not included since it, to certain extent, has similar meaning with the factor acceptance, one of the meaning-based coping behaviors. Among the meaning-based coping behaviours, letting go was integrated into the factor of non-reactivity. There were seven factors in total for meaning-based behaviours: present focus, acceptance, non-judgmental, describing, awareness, non-reactivity and beginner's mind.

Means and standard deviations were calculated for all of the coping strategies (Table 6.1). Of the three main strategies, construction personnel rated problem-based coping the highest, with a mean of 4.884, whereas they rated emotion-based coping the lowest (3.026). Of all 15 coping behaviours, *instrumental support seeking* had the highest mean (5.131). The second highest mean was for *planful problem solving* (4.993) followed by *positive reappraisal* (4.945). *Self-control* was rated the highest in the category of emotion-based coping strategies. The Cronbach's alpha values for all of the coping strategies were reasonable, ranging between 0.519 and 0.947 (Hair et al. 1998).

6.5.1.2 *Coping Behaviours of Construction Personnel by Age*

In order to investigate the impact of age on the coping behaviours of construction personnel, a one-way between-groups analysis of variance (ANOVA) was conducted. The results are shown in Table 6.2, Figure 6.2, Figure 6.3 and Figure 6.4. Personnel were divided into four groups according to their age (group 1: 29 years or less; group 2: 30–39 years; group 3: 40–49 years; group 4: 50 years and above). Among the respondents, 23.2% were aged 29 or lower, 25.9% were aged 30–39, 34.3% were aged 40–49 and 16.1% were aged 50 or above. Six of the 15 coping strategies had statistically significant differences at the $p < 0.05$ level for the four age groups: planful problem solving ($F = 7.998$, $p = 0.000$), instrumental support seeking ($F = 2.854$, $p = 0.037$), escapism ($F = 7.599$, $p = 0.001$), present focus ($F = 2.888$, $p = 0.040$), awareness ($F = 2.940$, $p = 0.038$) and beginner's mind ($F = 3.386$, $p = 0.022$).

To further compare the differences in coping behaviours between age groups, post-hoc comparisons were conducted using the *Tukey HSD test* for the six coping behaviours (Table 6.2). The results show that, for problem-based coping behaviours, the mean score of the coping behaviour of *planful problem solving* for group 1 ($M = 4.484$, $SD = 0.916$) was significantly different from all other three age groups: group 2 $M = 5.107$, $SD = 0.991$) with a mean difference of −0.623 ($p = 0.001$); group 3 ($M = 5.129$, $SD = 0.990$) with a mean difference of −0.645 ($p = 0.000$); and group 4 ($M = 5.243$, $SD = 0.760$) with a mean difference of −0.759 ($p = 0.001$). Groups 2, 3 and 4 did not differ significantly from each other in their mean planful problem solving. On the other hand, group 1 ($M = 4.839$, $SD = 0.932$) used *instrumental support seeking* significantly less than group 2 ($M = 5.277$, $SD = 0.906$) with a mean difference of −0.438 ($p = 0.047$).

Under the emotion-based coping behaviours, group 1 ($M = 3.433$, $SD = 1.208$) used *escapism* significantly more than group 3 ($M = 1.983$, $SD = 0.818$) with a mean difference of 1.450 ($p = 0.000$). Similarly, group 2 ($M = 2.638$, $SD = 1.368$)

Table 6.1 Coping behaviours of construction personnel.

Coping behaviours	Sample size	Mean score	SD	Alpha
Problem-based (mean = 4.884)				
1. Planful problem solving	277	**4.993**	0.980	0.717
– I made a plan of action and followed it				
– I came up with different solutions to the problems (e.g. time management, active responding)				
2. Instrumental support seeking	358	**5.131**	1.051	0.806
– I talked to someone who could do something concrete about the problem				
– I asked people who had had similar experiences what they did				
3. Confrontive coping	139	4.465	0.831	0.559
– I stood my ground and fought for what I wanted				
– I tried to get the person responsible to change his or her mind				
– I expressed anger to the person who caused the problem				
4. Positive reappraisal	139	**4.945**	0.796	0.595
– I came out of the experience better than when I went in				
– I changed or grew as a person in a good way				
– I rediscovered what is important in life				
Emotional-based (mean = 3.026)				
5. Emotional support seeking	281	3.836	1.351	0.787
– I tried to get emotional support from colleagues, friends or relatives				
– I talked to someone about how I feel				
6. Emotional discharge	68	2.831	1.465	0.542
– I try to reduce tension by eating more				
– I tried to lose myself for a while by smoking				
– I try to reduce tension by drinking more				
– I try to reduce tension by doing more physical exercise				
7. Escapism	137	2.412	1.211	0.670
– I escaped from the problem by doing unrelated things				
– I avoided phone calls to get rid of the problem				

Meaning-based (mean = 4.350)

	N	Mean		
8. Self-controlling	138	**4.368**	0.932	0.519
– I tried to keep my feelings to myself				
– I kept others from knowing how bad things were				
– I tried not to burn my bridges, but leave things open somewhat				
9. Present focus	90	4.581	0.318	0.532
– I feel connected to my experience in the here-and-now				
– I find myself preoccupied with the future or the past				
– I am able to focus on the present moment				
10. Acceptance	90	**4.752**	0.372	0.687
– I accept myself as I am				
– I am able to accept the thoughts and feelings I have				
– I can accept things I cannot change				
11. Non-judgmental	90	4.100	0.471	0.806
– I tell myself that I shouldn't be thinking the way I'm thinking				
– I tell myself that I shouldn't be feeling the way I'm feeling				
– I believe some of my thoughts are abnormal or bad and I shouldn't think that way				
12. Describing	90	4.096	0.569	0.947
– I'm good at finding the words to describe my feelings				
– I can easily put my beliefs, opinions and expectations into words				
– My natural tendency is to put my experiences into words				
13. Awareness	90	4.200	0.412	0.563
– I don't pay attention to what I'm doing because I'm daydreaming, worrying, or otherwise distracted				
– It is easy for me to concentrate on what I am doing				
– It seems I am "running on automatic" without much awareness of what I'm doing				
14. Non-reactivity	90	4.474	0.414	0.760
– When I have distressing thoughts or images I am able just to notice them without reacting				
– When I have distressing thoughts or images, I "step back" and am aware of the thought or image without getting taken over by it				
– When I have distressing thoughts or images, I just notice them and let them go				
15. Beginner's mind	90	4.244	0.419	0.778
– I was curious about my reactions to things				
– I was curious about what I might learn about myself by taking notice of how I react to certain thoughts, feelings or sensations				
– I was curious to see what my mind was up to from moment to moment				

Note: **Bolded** figure – highest rating amongst the 15 coping behaviors

Table 6.2 One-way between-groups ANOVA for coping behaviours of construction personnel in different age groups.

Coping behaviours	Age	Mean	SD	F (ANOVA)	Sig. (ANOVA)	Sig. (Levene)	Age group	Mean diff.	SE	Sig.
				One-way between-groups ANOVA			Post hoc test for coping behaviors with significant diff. scores			
Problem-based										
1. Planful problem Solving	**≤ 29**	**4.484**	**0.916**	**7.998**	**0.000**	**1.380**	**30–39**	**−0.623**	**0.162**	**0.001**
							40–49	**−0.645**	**0.152**	**0.000**
							≥ 50	**−0.759**	**0.196**	**0.001**
	30–39	5.107	0.991							
	40–49	5.129	0.990							
	≥ 50	5.243	0.760							
2. Instrumental support seeking	**≤ 29**	**4.839**	**0.932**	**2.854**	**0.037**	**1.722**	**30–39**	**−0.438**	**0.168**	**0.047**
	30–39	5.277	0.906							
	40–49	5.108	1.156							
	≥ 50	5.316	1.069							
3. Confrontive coping	≤ 29	4.306	0.746	0.947	0.420	1.026				
	30–39	4.533	0.663							
	40–49	4.583	1.015							
	≥ 50	4.500	0.894							
4. Positive reappraisal	≤ 29	4.799	0.737	1.202	0.312	2.938				
	30–39	4.922	0.575							
	40–49	5.049	0.993							
	≥ 50	5.167	0.789							
Emotion-based										
5. Emotional support seeking	≤ 29	3.567	1.449	1.677	0.172	0.771				
	30–39	4.100	1.309							
	40–49	3.754	1.391							
	≥ 50	3.769	1.186							

	Age	Mean	SD							
6. Emotional discharge	≤ 29	3.180	1.199							
	30–39	2.951	0.928							
	40–49	2.606	1.037	1.119	0.348	0.420				
	≥ 50	3.167	0.753							
7. Escapism	**≤ 29**	**3.433**	**1.208**							
	30–39	**2.638**	**1.368**	**7.599**	**0.000**	**6.942**	**40–49**	**1.450**	**0.326**	**0.000**
	40–49	1.983	0.818				**40–49**	**0.654**	**0.231**	**0.027**
	≥ 50	2.548	1.331							
8. Self-control	≤ 29	4.438	0.909							
	30–39	4.271	0.869	0.216	0.885	0.811				
	40–49	4.378	1.011							
	≥ 50	4.344	0.995							
Meaning-based										
9. Present focus	**≤ 29**	**4.258**	**0.952**							
	30–39	4.813	0.883	**2.888**	**0.040**	**0.163**	**40–49**	**−0.667**	**0.248**	**0.042**
	40–49	4.925	0.803							
	≥ 50	4.554	0.772							
10. Acceptance	≤ 29	4.484	1.108							
	30–39	4.479	0.709	2.425	0.071	0.882				
	40–49	5.150	1.084							
	≥ 50	4.957	1.012							
11. Non-judgmental	≤ 29	4.376	1.264							
	30–39	4.396	1.097	1.815	0.150	0.395				
	40–49	3.600	1.489							
	≥ 50	3.957	1.312							
12. Describing	≤ 29	4.269	1.676							
	30–39	4.333	1.393	0.532	0.661	0.350				
	40–49	3.983	1.652							
	≥ 50	3.797	1.617							
13. Awareness	**≤ 29**	**3.785**	**1.117**	**2.940**	**0.038**	**1.684**	**≥ 50**	**−0.911**	**0.308**	**0.021**
	30–39	4.229	0.841							

(Continued)

Table 6.2 (Continued)

Coping behaviours	Age	Mean	SD	F (ANOVA)	Sig. (ANOVA)	Sig. (Levene)	Age group	Mean diff.	SE	Sig.
				One-way between-groups ANOVA			Post hoc test for coping behaviors with significant diff. scores			
14. Non-reactivity	40–49	4.250	1.328							
	≥ 50	4.696	1.091							
	≤ 29	4.312	1.138	1.531	0.212	1.184				
	30–39	4.833	1.068							
	40–49	4.750	0.917							
	≥ 50	4.203	1.384							
15. Beginner's mind	**≤ 29**	**4.645**	**1.040**	**3.386**	**0.022**	**0.592**	**40–49**	**0.862**	**0.324**	**0.045**
	30–39	4.521	0.950							
	40–49	3.783	1.339							
	≥ 50	3.913	1.164							

Note: **Bolded** items - significant between-group differences

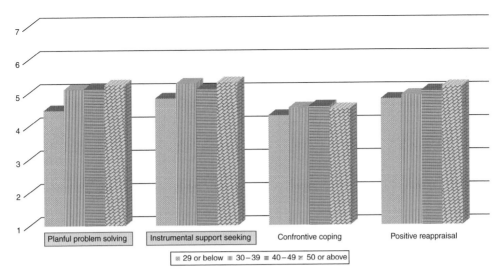

Figure 6.2 Problem-based coping behaviours used by construction personnel in different age groups. Note: Shaded items - Significant differences revealed in the one-way ANOVA.

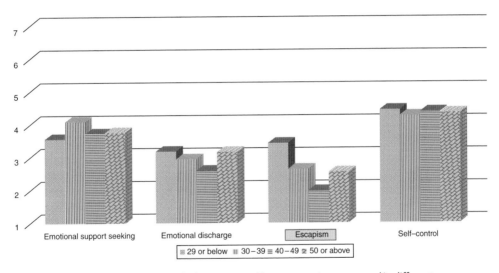

Figure 6.3 Emotion-based coping behaviours used by construction personnel in different age groups. Note: Shaded items - Significant differences revealed in the one-way ANOVA.

used significantly more *escapism* than group 3, with a mean difference of 0.654 (p=0.027).

In terms of meaning-based coping behaviours, construction personnel in group 1 used *present focus* (M=4.258, SD=0.952) significantly less than did those in group 3 (M=4.925, SD=0.803), with a mean difference of –0.667 (p=0.042); and construction personnel in group 1 used *beginner's mind*

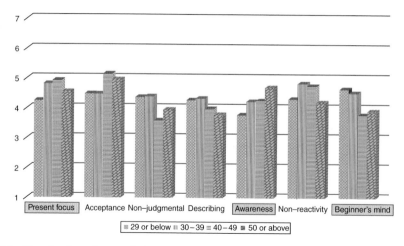

Figure 6.4   Meaning-based coping behaviours used by construction personnel in different age groups.
Note: Shaded items - Significant differences revealed in the one-way ANOVA.

(M=4.645, SD=1.040) significantly more than did those in group 3 (M=3.783, SD=1.339), with a mean difference of 0.862 (p=0.045). Furthermore, group 1 used the coping behaviour of *awareness* (M=3.785, SD=1.117) significantly differently from group 4 (M=4.696, SD=1.091). There were no significant differences in the use of other kinds of meaning-based coping behaviours between the four age groups.

6.5.1.3   Coping Behaviours of Construction Personnel by Gender

To compare the difference of coping behaviours between male and female construction personnel, one-way analysis of variance was conducted (see Table 6.3, Figure 6.5, 6.6 and 6.7). It is interesting to note that there was no significant difference between males and females, except for *confrontive coping* and *emotional support seeking*. Male construction personnel reported more confrontive coping than female construction personnel, with a mean difference of 0.428 (F=4.258, p=0.041). In contrast, male construction personnel reported *less* emotional support seeking than female construction personnel, with a mean difference of –0.686 (F=10.439, p=0.001).

6.5.1.4   Relationships Between Coping Behaviours, Stress and Performance of Construction Personnel

Pearson correlation is used to investigate the interrelationships between stress, coping behaviours and performance. Table 6.4 shows the correlations between the 24 variables for the construction personnel. Problem-based coping behaviours were positively related to performance: (1) planful problem

Table 6.3   One-way between-groups ANOVA for coping behaviours of construction personnel of different genders.

Coping behaviours	Gender	Mean	SD	F (ANOVA)	Sig. (ANOVA)	Sig. (Levene)
Problem-Based						
1. Planful problem solving	Male	5.023	0.986	0.482	0.488	0.028
	Female	4.900	0.938			
2. Instrumental support seeking	Male	5.149	1.053	0.351	0.554	0.363
	Female	5.041	1.083			
3. Confrontive coping	**Male**	**4.521**	**0.830**	**4.258**	**0.041**	**1.018**
	Female	**4.093**	**0.757**			
4. Positive reappraisal	Male	4.989	0.807	2.912	0.090	0.084
	Female	4.648	0.661			
Emotion-based						
5. Emotional support seeking	**Male**	**3.731**	**1.360**	**10.439**	**0.001**	**2.468**
	Female	**4.417**	**1.209**			
6. Emotional discharge	Male	2.983	0.980	0.542	0.464	0.317
	Female	2.800	1.056			
7. Escapism	Male	2.364	1.170	0.985	0.323	4.234
	Female	2.677	1.468			
8. Self-control	Male	4.344	0.927	0.606	0.438	0.954
	Female	4.528	0.977			
Meaning-based						
9. Present focus	Male	4.573	0.909	0.150	0.700	2.652
	Female	4.750	0.354			
10. Acceptance	Male	4.729	1.057	0.955	0.331	1.801
	Female	5.250	0.500			
11. Non-judgmental	Male	4.097	1.347	0.011	0.918	2.106
	Female	4.167	0.577			
12. Describing	Male	4.112	1.609	0.195	0.660	0.913
	Female	3.750	1.101			
13. Awareness	Male	4.217	1.159	0.419	0.519	0.022
	Female	3.833	1.171			
14. Non-reactivity	Male	4.457	1.185	.397	0.530	3.189
	Female	4.833	0.333			
15. Beginner's mind	Male	4.236	1.200	0.089	0.766	2.995
	Female	4.417	0.319			

Note: **Bolded** items - significant between-group difference

solving (C1) and instrumental support seeking (C2) were positively related to personal satisfaction (PS: 0.381, 0.299, $p < 0.01$), interpersonal relationships (IR: 0.336, 0.269, $p < 0.01$), project outcomes (PO: 0.408, 0.177, $p < 0.01$) and sense of belonging (Belong: 0.137, 0.136, $p < 0.05$) and negatively related to intention to stay (IS: −0.200, −0.216, $p < 0.01$); (2) both confrontive coping (C3) and positive reappraisal (C4) were positively associated with personal satisfaction (PS: 0.200, $p < 0.05$; 0.435, $p < 0.01$) and project outcomes (PO: 0.229, $p < 0.05$; 0.268, $p < 0.05$); and (3) positive

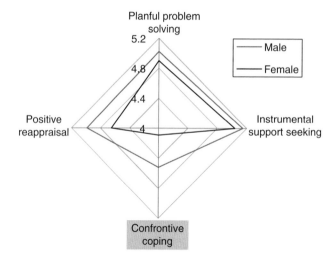

Figure 6.5 Problem-based coping behaviours used by construction personnel of different genders. Note: Shaded item - Significant differences revealed in a one-way ANOVA.

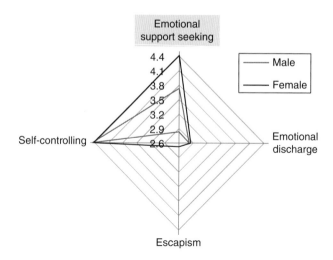

Figure 6.6 Emotion-based coping behaviours used by construction personnel of different genders. Note: Shaded item - Significant differences revealed in a one-way ANOVA.

reappraisal was also negatively related to intention to stay (IS: –0.259, $p < 0.01$). Instrumental support seeking (C2) was negatively related to physical stress (PS: –0.129, $p < 0.05$), but confrontive coping (C3) was positively related to physical stress (PS: 0.274, $p < 0.01$) and emotional stress (ES: 0.200, $p < 0.05$).

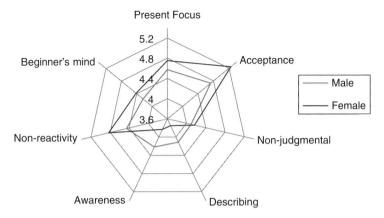

Figure 6.7 Meaning-based coping behaviours used by construction personnel of different genders. Note: There were no significant differences between males and females for meaning-based coping behaviours.

None of the emotion-based coping behaviours were significantly related to any of the six types of performance, except that emotional discharge (C6) was related to personal satisfaction (PS: –0.258, p < 0.05). Similarly, emotional support seeking (C5) had no significant relationship to any kind of stress. Escapism as a coping behaviour (C7) was significantly related to all three kinds of stress: work stress (WS: 0.236, p < 0.01), emotional stress (ES: 0.170, p < 0.05) and physical stress (PS: 0.353, p < 0.05). Emotional discharge (C6) was related to physical stress (0.228, p < 0.05) and self-controlling coping behaviour (C8) was correlated with emotional stress (0.217, p < 0.05).

The results indicated that describing (C12) was related to the first three types of performance: personal satisfaction (PS: 0.213, p < 0.05), interpersonal relationships (IR: 0.504, p < 0.01) and project outcomes (PO: 0.210, p < 0.05). Awareness (C13) was related to another three types of performance: safety behaviour (Safety: 0.545, p < 0.01), sense of belonging (Belong: –0.327, p < 0.01) and intention to stay (IS: –0.264, p < 0.05). In addition, non-judgmental (C11) was significantly related to three types of performance: personal satisfaction (PS: –0.282, p < 0.01), sense of belonging (Belong: –0.252, p < 0.05) and intention to stay (IS: 0.238, p < 0.05). Meanwhile, acceptance (C10) only had a significant negative relationship with intention to stay (IS: –0.253, p < 0.05) and beginner's mind (C15) only had a significant positive relationship with interpersonal relationships (IR: 0.293, p < 0.01). At the same time, three meaning-based coping behaviours were significantly related to stress: present focus (C9) was related to work stress (WS: –0.292, p < 0.01) and emotional stress (ES: –0.319, p < 0.01); non-judgmental (C11) was related to physical stress (PS: –0.234, p < 0.05); and awareness was related to emotional (–0.403, p < 0.01) and physical stress (–0.543, p < 0.01).

Table 6.4 Correlations between coping behaviours, stress and performance.

Factors		Stress			Performance					
		WS	ES	PS	PS	IR	PO	Safety	Belong	IS
C1	Planful problem solving	0.001	0.027	-0.115	**0.381****	**0.336***	**0.408****	-	**0.137***	**-0.200****
C2	Instrumental support seeking	0.046	0.053	**-0.129***	**0.299****	**0.269****	**0.177****	-	**0.136***	**-0.216****
C3	Confrontive coping	0.140	**0.274****	**0.200***	**0.200***	0.038	**0.229***	-	-0.031	-0.104
C4	Positive reappraisal	-0.019	0.141	-0.044	**0.435****	-	**0.268****	-	0.154	**-0.259****
C5	Emotional support seeking	0.059	-0.023	0.051	0.024	0.024	-0.047	-	-0.007	-0.020
C6	Emotional discharge	-0.007	0.220	**0.262***	**-0.258***	-	-	-	-0.041	0.081
C7	Escapism	**0.236****	**0.170***	**0.353***	-0.087	-0.050	-	-	0.023	0.019
C8	Self-control	-0.030	**0.217***	0.135	0.112	0.061	0.154	-	0.011	0.120
C9	Present focus	**-0.292****	**-0.319****	-0.198	-0.066	0.120	0.127	0.155	-0.207	0.173
C10	Acceptance	-0.189	-0.092	-0.166	-0.166	0.010	-0.022	0.165	-0.204	**-0.253***
C11	Non-judgmental	-0.035	-0.120	**-0.234***	**-0.282****	-0.036	-0.168	0.152	**-0.252***	**0.238***
C12	Describing	-0.183	-0.116	0.083	**0.213***	**0.504****	**0.210***	-0.167	-0.111	0.019
C13	Awareness	-0.078	**-0.403****	**-0.543****	-0.136	0.119	0.103	**0.545*****	**0.327****	**-0.264***
C14	Non-reactivity	-0.037	-0.012	0.126	0.138	0.095	0.081	-0.067	0.052	0.043
C15	Beginner's mind	0.033	-0.051	-0.007	0.062	**0.293****	0.006	-0.205	-0.055	-0.068

Note: **Bolded** figures - significant correlation coefficients

*correlation is significant at the 0.05 level (two-tailed)

**correlation is significant at the 0.01 level (two-tailed).

WS = work stress; ES = emotional stress; PS = physical stress; PS = personal satisfaction; IR = interpersonal relationships; PO = project outcomes; Safety = safety behaviours; Belong = sense of belonging; IS = intention to stay (see Section 5.5.1 for details on these variables).

6.6   Case Studies

6.6.1   Drainage Installation in a City Centre

This project, which was located in a city centre between two busy main roads, aimed to install new underground pipework and modify the existing pipes along the street. Before laying and fixing a new drainage system, a temporary diversion sewage network had firstly to be installed. A reconnection of the drainage system was then required in order to maintain regular functions for local users. The total cost of this drainage work was US$5 million over a construction period of 200 calendar days. Multiple stakeholders were involved in and affected by the project, including government departments (the Civil Engineering Development, the Drainage Services, Police and Water Supply Departments and the Urban Renewal Authority), the District Council and local residents. Hence, coordination among these different departments and other project participants such as the clients, consultants and representatives of the public was critical for successful implementation.

The installation of new drainage underneath the road and the diversion of existing drainage system were both difficult tasks, particularly in the busy local environment. The most complicated and difficult aspects were getting the cooperation of nearby residents during the suspension of the sewage system and applying the procedure for obtaining work permissions on the existing road. At the beginning of the project, the project manager outlined a programme including a start date, milestone dates for the application of various permissions from various government departments and the completion dates for the installation and diversion of the drainage system. During the construction period, the project manager and engineer regularly met with staff of the different government departments, public representatives, consultants and the client. The engineer was exhausted because he often worked overtime in order to prepare and arrange for these meetings. He said that he wanted to employ an assistant, but it was hard to recruit one due to the limited career prospects on offer available in this project.

Moreover, the drainage project was also expected to have a reduced impact on local residents. The project manager therefore had to organise internal meetings with subcontractors and external meetings with different government departments and public representatives to agree a more effective construction method and procedure. The goal of this was to minimise inconvenience to end users, be they local residents, shopkeepers, commercial property owners, or bus and taxi drivers. The representatives of the public complained regularly about the slow progress of the project and its continuous interference with their interests. They thought that the contractor was allocating insufficient staffing resources to the work and proposed a number of ways to speed up the work and minimise the disturbance. The consultant made no comment on these suggestions, but the project manager invited the public representatives to clarify their proposals. However, the representatives had assumed that different ideas would be put forward by the consultants and the project manager, since they were the ones with the relevant professional expertise. Moreover, they were also concerned about site safety and suggested closing the entrance gate to avoid incidents involving passers-by entering the site. In fact, the project manager received a number of complaints from local residents for not providing adequate safety measures, especially warning signs at night.

Due to the volume of complaints received, the site planning and the project programme changed frequently, which significantly increased the workload of the construction personnel. The project manager often had to deal with the reaction of workers who expressed their unwillingness to comply with orders to work late. Those staff then had less time to enjoy dinner in the evening with friends, while the project manager also had to stay late in the office and had limited time for his family.

To deal with their stress, these construction personnel were more likely to use *problem-based coping behaviours* and attempt to solve the problem directly (see Table 6.5). As the drainage project was located near the local community, the engineer often carried out site inspections and proposed several construction methods to the project manager for minimising the impact on local residents. When receiving complaints from the public, the project manager also tried to think of 'a couple of practical solutions to meet their requirements'. If necessary, he would discuss with fellow project managers in other construction companies, so as to exchange experience in handling issues involving the public. Moreover, the permission of government departments could influence the whole project duration and implementation. To successfully obtain such permission, the engineer needed to work overtime and prepare all the necessary documents in advance to a high standard. Although the workload was heavy, he thought that he had 'learned a lot from the preparation of the documents and the problem solving'. In contrast, the construction workers preferred to adopt confrontive coping behaviours to deal with their stress. For example, one worker said that he had spoken to the foreman to complain about the ever-changing construction plan. These construction personnel considered that the adoption of problem-based coping could improve their effectiveness and ultimately the delivery of the project.

Although these staff mainly used problem-based coping behaviours, the project manager and the construction worker also applied *emotion-based coping*. As well as discussing problems with friends, the project manager also sought their emotional support. On the other hand, the construction worker also said that he felt exhausted because of the late orders and overtime. He often complained about his heavy workload to his friends and drank a lot of beer to make him feel better.

These construction personnel also applied several *meaning-based coping behaviours*, such as present focus, remaining non-judgmental and raising awareness. The project manager mentioned that he tried to 'focus on the present moment and accept the current situation' when facing serious problems (such as public disturbance and project delays) and suffering from stress. The engineer also stated that he tried to listen to the public representatives' complaints in a non-judgmental way at their regular meetings. Afterwards, he discussed their views with other project members and made as much effort

Table 6.5 Coping behaviours used by construction personnel involved in the drainage project.

Coping behaviours	Construction personnel
Problem based	PM: Our project team often meets with public representatives to understand their needs and manage their influence over the drainage project. When receiving public complaints, I usually firstly think **about a couple of practical solutions** to meet their requirements … Sometimes it is difficult to meet the requests of local residents. If I think that I cannot deal with the problems by myself, I prefer to **discuss them with fellow project managers** in other construction companies. Eng: To get the permissions for the construction, I often need to prepare documents for meetings with government departments. These are time-consuming tasks. However, when we finally get the permission, I usually think that I have **learned a lot from** the preparation of documents and the problem solving. CW : I don't understand why the construction plan is always changing in this project … The last late orders we received were really enough. I **spoke to the foreman** and complained about this. He said that he had already **passed our complaints to the project manager.**
Emotion based	PM : I sometimes feel upset about the nonsense public complaints (such as closing the entrance gate all the time). I like to go for a smoke on my own and phone my friends to **seek emotional support**. CW : I often receive late orders from the project manager and work overtime. I feel exhausted after a long working day. I often **complain about the heavy workload and drink a lot of beer.**
Meaning based	PM : When I am worried about the project (such as public disturbance, which would delay the whole construction process), I can sense my nervous feelings. I often try to **focus on the present moment and accept the current situation**. Then, I try to accept the things I cannot change right now before discussing the problems with my subordinates. Eng : When some of the public representatives express their ideas about the project and some of them are unfriendly, I normally tell myself to listen to their opinions in a **non-judgmental** way first … This can help me in making a fair judgment and response. CW : The project manager always reminds us the importance of safety. Hence, I will **pay attention** to all safety precaution measures and regulations in detail until my tasks are completed properly.

PM = Project manager; Eng = Engineer; CW = Construction worker.

as possible to meet the public's requirements. Moreover, one of the construction workers also confirmed that he was aware of the safety precautions and equipment and paid attention to these until he had completed his tasks.

6.6.2   Luxury Low-Density Residential Building

This residential building project was initiated by a large private developer. It set out to construct luxury low-density residential buildings and provide over 300 four-storey houses. The overall project duration was 36 months and the contract sum for the substructure project was US$55 million. As a large project, various parties were involved including a project manager, architect, engineer, quantity surveyor, sales departments, main contractor and many subcontractors. To facilitate communication among different parties and solve project problems, the developer conducted three systematic value management (VM) workshops to make critical decisions.

At the predesign stage, the developer carried out a VM workshop to discuss the development strategy, which aimed to establish high-quality luxury residential buildings. In the substructure tendering period, the developer received the tender documents before the master plan had been confirmed. The lump sum contract with the specific schedule of rates had been fixed, while the development design had not yet been finalised. After the project team had finalised the soil investigation report, the master plan was revised, which induced lots of conflict among the different parties. Due to the amended foundations, the actual construction period, particularly for the foundation contract, was extended. The developer therefore held a further VM workshop involving multiple parties to discuss the situation and solve the problems arising from it. The project team suggested several methods to safeguard the developer's profit margin, including (i) the identification of subsequent construction phases; (ii) the reduction of the floor area; and (iii) the revision of the development value and construction costs. After this workshop, the participants agreed to change the design from four- to two- and three-storey houses.

In the substructure construction process, more conflict arose between the parties over issues including the underground drainage, the combination of construction phases and the provision of sea views. The overall construction progress was also delayed due to late action by the engineers. Another VM workshop was then conducted with internal parties, external professionals and the main contractor and subcontractors. The architect then finalised the development design, while the engineers specified the type of foundation and site formation. The quantity surveyors re-estimated the project sums and stated specific qualifications in the contract documents. Based on the site environment and updated information, the project manager decided to combine the construction of two of the phases and asked the main contractor to propose an updated project programme for agreement.

The overall construction period and sequences were then fixed. However, the consultants did not implement the schedule and submit working drawings on time, which significantly affected the whole progress of the project. As the project was so far behind schedule, the participants tried to figure out solutions in the next VM workshop. They rescheduled the construction programme and revised their working sequences in order to shorten the construction period and meet the original completion date. However, such 'rush' progress definitely decreased project quality.

In reality, the completion date of each project phrase remained behind schedule for each milestone. The main contractor had an opportunity to reclaim its losses and expenses due to the lateness of the working drawings issued by the structural engineer. Moreover, lots of problems arose among multiple parties involved in the substructure construction process. For example, the project manager was dissatisfied with the design proposal submitted by the engineers and the architect could not fix the detailed design. Hence, a third VM workshop was carried out in the construction phase. Although a programme was finally specified, that workshop was still considered to have failed to reach a satisfactory outcome.

When facing stress, the construction personnel (project manager and quantity surveyor) involved with the substructure project used *problem-based coping behaviours* (such as planful problem solving and positive reappraisal) to address the problems directly via the VM workshop and manage the sources of their stress (see Table 6.6). The problems with this project affected the overall progress of the construction and caused stress to the project manager. In order to address the problems he was facing, the project manager kept up pressure on the engineer to provide drawings and discussed the problems with other stakeholders. The architect reported that he would often 'figure out a number of ideas, some of them even crazy', to design the building. Since the engineer had failed to submit his drawings on time, the whole construction progress was seriously delayed. Consequently, the engineer was criticised by other project team members and experienced great stress as a result. In fact, he had already tried to report his workload issue to his superiors and had asked for more staff to be assigned to the work. However, he received negative feedback. The engineer was unable to finish the work as scheduled, which subsequently affected the ability of the quantity surveyor to estimate the project accurately. However, on a positive note, the quantity surveyor believed that she had nevertheless 'learned a lot about team decision making from the VM workshop in this project'.

These construction personnel also used *emotion-based coping behaviours* to deal with stress, but the final project could still not be completed on budget and within the agreed schedule. When feeling upset about the conflicts among different parties, the project manager liked to drink and smoke a lot, which affected his physical health, while the engineer escaped from his stress by 'imagining that the late submission of drawings is not a big problem'. However, his failure to submit the drawings actually resulted in delays to the whole project, which in reality created more stress. Furthermore, the late submission of drawings influenced the activities of the construction workers on the site. Since the worker interviewed considered the structural engineer to be an internal professional employed by the client, he controlled his feelings and tried not to be anxious or angry with him even though he had to work extremely late hours as a result.

On the other hand, some construction personnel involved in this project also mentioned the application of *meaning-based coping behaviours*. The project manager listened to others' opinions and made no judgements when participating in the VM workshops. This non-judgemental attitude helped him maintain good interpersonal relationships with multiple parties. The architect also described how, because of his knowledge of the site environment, 'it was easy for me to describe the problems in the workshops'. Given his descriptions, other participants understood the situation better and were able to come up with several feasible solutions. Moreover, present focus was also used by the construction worker. He agreed that the tight project duration caused him stress. To manage this, he tried to 'focus on the present work tasks first'. He believed that this helped him to accept the current situation more easily.

Table 6.6 Coping behaviours used by construction personnel in the luxury homes project.

Coping behaviours	Construction personnel
Problem based	PM: The IA project member did not work together in the team. He always failed to submit drawings on time for estimation. This is the key problem in this project that I have to face, so I keep **forcing him to provide the drawings** and **discuss this issue with other members** for better solutions. Arch: Normally, I will **figure out a number of ideas**, some of them even crazy, to design the building. Some of these ideas were actually borrowed from my previous experience. Eng: We set out the schedule very clearly, but I was really too busy at that time and could not finish the work as set . . . I tried to **report my difficulties to my supervisor and asked for human resources rearrangement**. The feedback was that 'We can provide no technical support, sorry'. To be honest, the criticisms from other project members placed me under great stress. QS: It is impossible for us to estimate the project budget accurately if the drawings, particularly the structural drawings, are not finalised. However, I still **learned a lot about team decision making from the VM workshop** in this project.
Emotional based	PM: I feel upset due to the conflict between different parties. I often **drink and smoke a lot** to relieve my tension. However, my physical health has been affected by my bad habits. I easily get colds and coughs. Eng: The site foundation is different from what we expected. We needed to revise the drawings based on the site foundation. As the project duration is tight, I feel so stressed. I **often imagine that it will not be a big problem** if I submit the updated drawings several days late. In the end, I found out that the progress of the substructure project was far behind schedule, which made me even more anxious and crazy. QS: I was requested to re-estimate the project sum many times. I am stressed because of the increased workload. I need to **drink a lot of coffee** when I work overtime. However, that increases sleep disorders, which in turn affect my work performance. CW: Our work was significantly influenced by the late submission of structural drawings. As the structural engineers are internal professionals employed by the developer, I often try to **keep my feelings to myself** and not be anxious or angry. . . . However, we found that the completion date was much later than the original expectations. I feel so stressed by the tight project progress.
Meaning based	PM: In the VM workshop, I listened to other opinions with an open mind and **made no judgement** on all opinions. This non-judgemental attitude helped me maintain a good relationship with both internal and external consultants as well as the main contractor and subcontractors. Arch: I carried out site inspections frequently and was familiar with the site environment. It was easy for me to **describe the problems in the workshop**. My description and explanation helped the participants understand the actual situation and propose solutions. CW: The project schedule is very tight. I often feel anxious about the project progress. To manage my stress, I tried to **focus on the present work tasks** first . . . so that I realised that I am making some progress at least, making it easier to **accept the current situation** of the problems. . . . When I encounter some technical problems, I don't mind sharing them with some of the 'green' workers. Although they are not as experienced as I am, they can sometimes provide me with inspiration due to their **beginner's mind**. From them, I have learned the importance of beginner's mind.

PM = Project manager; Arch = Architect; Eng = Engineer; QS = Quantity surveyor; CW = Construction worker.

6.7   Discussion

6.7.1   Coping Behaviours of Construction Personnel

Construction personnel have to equip themselves with the knowledge to solve all sorts of technical and management problems in their daily work. Therefore, it is not surprising to find that they like to use problem-based coping behaviours such as *instrumental support seeking*, *planful problem solving* and *positive reappraisal* to relieve stress (Table 6.1).

Construction projects are often complicated and involve diverse professional tasks such as architectural and structural design, building services engineering, contractual procurement process, budget estimation and construction programming, which are all interrelated in a complex system. Moreover, change is constant in any construction project. Under such circumstances, construction personnel have to plan ahead and prepare several solutions to expected problems so that the construction process will not be affected. It is impossible for construction personnel to solve all problems by themselves, especially personnel who participate in mega projects in large cities. Therefore, construction personnel need to *seek instrumental support* to deal with stressful problems, which is the most efficient way of alleviating stress (Leung, Wong and Oloke 2003). For example, they can seek innovative geo-technological support for unexpected soil conditions, structural elements for particular architectural design, detailed curtain wall design for budget estimation and extra-manpower for demanding workload. Of course, construction personnel also need to *reappraise* the stressor and *plan* alternative possible methods.

In order to complete construction projects successfully, the efforts of every construction personnel are equally important. Cooperation and communication is therefore important. Hence, construction personnel tend to *control or hide* their emotions when they are facing particular stressful problems, so that communication channels can be kept intact (Folkman et al. 1987). Although an individual may feel better, in terms of personal face and ego, keeping negative feelings (e.g., anxiety, depression and anger) to oneself and preventing others from discovering their stressful situation can be detrimental for solving the problem.

Construction projects consist of a series of complicated processes. Therefore, construction personnel have to pay attention in each moment in order to implement the tasks as planned. Moreover, *acceptance* of the influence of stress towards oneself, in contrast with avoiding or escaping from it, can allow construction personnel to focus on the problems in advance. Since acceptance is a non-evaluative attitude towards self and others, which involves unconditional acceptance of one's self, one's own weakness, failures and unfortunate experiences (Buchheld, Grossman and Walach 2001), it will help construction personnel find effective ways to solve problems. As mentioned earlier, one of the main positive outcomes of meaning-based coping is that it sustains the adoption of other coping behaviors, adaptive coping in particular.

6.7.2   Coping Behaviours of Construction Personnel of Different Age Groups

As shown in Table 6.2, six of the 15 coping behaviours have significant differences between age groups. Construction personnel aged 29 or below rated *planful problem solving* significantly lower than the other three age groups. Since young construction personnel have just graduated from university and started working in the construction industry, they may be threatened by the stressors that they may not be able to handle and fail to plan before they act. As they work more years in the industry, their experiences can accumulate and help them cope with stressful problems. Then they will become more confident in dealing with these problems and they can come up with different solutions and make plans to implement them. In addition, more senior people generally have higher positions in their organisations and more influence on the working environment. Hence, they normally have a stronger belief that they can alter unfavourable working environments, which can increase their propensity to use the coping behaviour of problem planful solving. In fact, many previous studies have found that older adults usually select more effective coping strategies, such as planful problem solving, compared with younger adults (Blanchard-Fields, Chen and Norris 1997; Blanchard-Fields, Mienaltowski and Seay 2007; Watson and Blanchard-Fields 1998).

Similarly, young construction personnel (age 29 or below) are less likely to use the coping behaviours of *instrumental support seeking* to deal with stressful problems, compared with construction personnel aged between 30 and 39. This reflects that young construction personnel with limited working experience do not have the social networks to use to seek support for the difficulties in their work, while construction personnel aged 30 to 39 have established more comprehensive social networks to support their work. When they encounter problems, construction personnel agea 30 to 39 probably know who can provide help and seeking their support will be effective for them (Strough et al. 2008).

It was more common for construction personnel aged 29 or below to use *escapism*. Although construction personnel are equipped with a certain amount of knowledge from school, practical tasks are often different from what they learned in books and this causes difficulty. Therefore, it is not surprising to find that construction personnel aged 29 or below prefer escapism and passive dependence more than older groups (Blanchard-Fields, Mienaltowski and Seay 2007).

The escapism behaviours that young construction personnel use could also be explained by the meaning-based coping behaviours of *present focus* and *awareness* that older construction personnel use (between 40 and 49 and older than 50, respectively). Acting with awareness means focusing on one thing at a time and wholly participating in an activity with undivided attention (Baer, Smith and Allen 2004; Brown and Ryan 2003). Experienced construction personnel usually know what they can do and how they can do it. Therefore, it is easier for them to pay attention in each present moment and be aware of each task than the younger construction personnel.

It was also found that younger construction personnel tend to use beginner's mind more than the old groups (between 40 and 49). As beginners in the industry, young construction personnel have a lot of new things to learn. Hence, they are usually curious about everything in the workplace and prepared to learn (Bishop et al. 2004). In contrast, older construction personnel have a lot of experience in construction projects. It is difficult for them to have a beginner's mind to manage, evaluate a situation in a non-biased way (disregarding previous positive or negative experience) and implement their professional knowledge in a non-intuitive way. Hence, it is no surprise that construction personnel age 40 and above use beginner's mind significantly less than the young group.

6.7.3   Coping Behaviours of Construction Personnel of Different Genders Groups

The results in Table 6.3 show that male and female construction personnel have no significant differences in any of the meaning-based coping behaviours. Among the eight problem-based and emotion-based coping strategies, there were only significant differences in two coping behaviours: confrontive coping and emotional support seeking. *Confrontive coping* refers to aggressive, proactive actions in interpersonal relationships to handle problems (Folkman and Lazarus 1988). Since males usually prefer to take concrete actions versus attend to interpersonal emotional concerns (Vingerhoets and Van Heck 1990) and males are less avoidant of confrontations (Kenrick, Neuberg and Cialdini 2010), it is not surprising that they are more likely to use confrontive coping.

The results showed that female construction personnel were more interested in seeking emotional support than their male counterparts, which fits with previous studies of general adults (Sullivan 2002). Macho culture is prevalent in the construction industry and macho culture sees admitting to stress as a sign of weakness. This makes male construction personnel unwilling to admit their stress, refusing to ask for help in resolving it. In addition, men are socialised in ways that discourage them from seeking emotional support, but women are socialised in ways that encourage it (Stokes and Wilson 1984). Hence, compared with their female counterparts, it is very uncommon for male construction personnel to use emotional support seeking as a coping behaviour. Instead, they more often want to keep their stress a secret.

6.7.4   Coping Behaviours, Stress and Performance of Construction Personnel

As shown in Table 6.4, each problem-based coping behaviour (i.e., planful problem solving, instrumental support seeking, confrontive coping and positive reappraisal) can enhance at least two types of performance. This actually fits with previous studies, which found that problem-based coping behaviours are adaptive (Haynes and Love 2004; Leung, Liu and Wong 2006). People seem to be more likely to use *planful problem solving* in situations that are appraised as changeable or challenging and in work situations with established goals (Folkman et al. 1986; Oakland and Ostell, 1996). *Instrumental support seeking* involves seeking practical help with work tasks in the form of information, advice or assistance (Richards and Schat 2011). *Positive reappraisal* is often the first step towards a productive reengagement with the stressful event. On-site work in construction projects is always challenging because of the dynamic and complicated nature of construction projects (Baiden and Price 2011; Tanskanen et al. 2008). Thus, construction personnel often need to deal with problems instantly (e.g., the air-conditioning system has insufficient capacity for the revised layout in the design stage and there is a lack of materials and workers during the construction stage) in order to complete the tasks according to the schedule within the budget.

Construction personnel who use positive reappraisal may positively reappraise difficult work as an opportunity to learn and improve themselves, rather than as an instance of bad luck and chance to be criticised by their supervisor again. That sort of positive reappraisal can improve their job satisfaction and project outcomes. The more construction personnel use planful problem solving and instrumental support seeking, the more they deal with problems in the projects efficiently and positively (Poon, Lee and Ong 2012). This makes them more satisfied with their own performance, helps them develop closer relationships with colleagues and project team members (Karimzade and Besharat, 2011) and helps them achieve better project outcomes, a sense of belonging in the company and stronger intention to stay. Subsequently, there are fewer complaints or conflict with their supervisors, colleagues and subordinates because they can deal with tasks in advance. In addition, construction personnel using instrumental support seeking are less likely to suffer from physical stress (e.g., sleep disorders, headaches and back pain) (Owasoyo, Neri and Lamberth 1992).

Confrontive coping was found to be positively related to two kinds of stress: emotional stress and physical stress, as well as two types of performance: personal satisfaction and project outcomes. Due to the demanding tasks in daily work, construction personnel may express their anger and aggression to others who make mistakes (Howard and McKillen 1990; Lambert and Lambert 2008). For example, a project manager might show anger to a steel worker when that person makes mistakes in the steel work, or a building services engineer might blame an air-conditioning mechanical ventilation subcontractor for mixing up the installation with another subcontractor, all of which may cause delays in the whole project. The results of this study fit with previous studies, which found that confrontive coping is associated with demanding tasks, tends to increase negative emotions and is negatively correlated with psychological and physical health (Farhall and Gehrke 2011; Penley, Tomaka and Wiebe 2002; Whittington and Wykes 1994). Thus, construction personnel using this type of coping were found to have more emotional and physical stress. However, construction personnel using confrontive coping behaviour are likely to deal with problems aggressively with a huge amount of personal effort. This effort can help them solve the problems in the end, which is essential for their satisfaction and the project outcomes.

Construction personnel suffering from stress may use alcohol and cigarettes to discharge their negative emotion. However, emotional discharge cannot help construction personnel solve their actual problems, while it can damage their physical health (i.e., empirical studies have shown that drinking and smoking can cause headaches, stomach aches and so on; Carter and Chapman 2003; Rehm et al. 2003). With the stressful situation still unsolved after emotional discharge, construction personnel may feel more physically stressed when they go back to work and face the stressor, say after drinking. Moreover, it is hard for construction personnel with serious physical stress and poor health to improve their performance. Hence, emotional discharge is negatively related to satisfaction with personal performance.

Escapism is a distinct personal limitation that contributes to stress reactions beyond the influence of stressors and support in the organisational environment (Leiter 1991). Although construction personnel have the ability to deal with many things, such as a high task load and deadlines, they can only manage to cope with less than their actual ability. Escapism is an emotion-based factor that is related to negative outcomes, such as sleeping more than usual and avoiding other people (Aldwin and Revenson 1987). When construction professionals encounter task difficulties or demands that exceed their ability (i.e., objective stress) for a long time, their physical and psychological health are inevitably impaired with consequences such as headaches, back pain, depression and anxiety. Thus, it is no surprise that they prefer to escape stressful tasks (Latack 1986).

When people use *self-controlling* coping behaviour, they usually try to stay calm to deal with the problem and not express their emotions (French 2004). Although construction personnel who use self-controlling may not express anger or anxiety, they may inhibit their feelings and actions towards the stressor, focusing their attention on controlling their emotions. They may conceal the situation and their stress rather than managing and overcoming the problem. This can eventually make them emotionally drained or frustrated (Violanti 1993), which escalates their emotional stress in a vicious cycle.

Present focus was negatively related to work stress and emotional stress. Since construction personnel have a lot of work to do every day, especially when facing deadlines and milestones, they may have put a lot of attention into every task. With the help of the present-focus coping behaviours, construction personnel can pay attention to the here and now, which is helpful for dealing with their tasks efficiently. Hence, they can handle more tasks at the same time, minimise the gap between their actual ability and their expected ability, easily meet the project requirements and establish good relationships with colleagues and project team members. The more construction personnel use this coping strategy, they can accomplish more tasks while feeling less work stress and emotional stress.

Acceptance is defined as being experientially open to the reality of the present moment (Roemer and Orsillo 2002). It involves a conscious decision to abandon one's agenda to have a different experience and instead hold an active process of 'allowing' current thoughts, feelings and sensations (Hayes, Strosahl and Wilson 1999). Hence, it is an active process in that the construction personnel choose to take what is offered with an attitude of openness and receptivity to whatever happens to occur in the field of awareness. Construction personnel using this coping behaviour may opt to follow their own heart, which could explain why they reported lower intention to stay in this study. Thus, acceptance may be related to turnover for people working in stressful environments.

Non-judgemental is another meaning-based coping behaviour and it refers to facing difficulties without jumping to conclusions about the value of those occurrences (Kabat-Zinn 1993). When construction personnel do not use judgement to manage their daily work, they are more likely to be flexible in accepting change. This would reduce the extent and time of exposure of construction personnel to physical stress. However, it is difficult for construction personnel to handle complex project problems with a non-judgemental attitude. In this case, it is harder for construction personnel who use non-judgemental coping behaviours to improve their performance and hence their satisfaction with their jobs. It is also a negative coping behaviour for organisational performance, in terms of a lower sense of belonging, although it is associated with higher intention to stay.

Describing is using words and mindfulness to express one's feelings and it is likely to be associated with more complex descriptions of one's thoughts as contextual, relativistic, transient and subjective (Segal, Williams and Teasdale 2002). Construction personnel using this coping behaviour can easily describe their feelings, experiences, beliefs and expectations. In general, describing takes account of cognitions, affect and somatic sensations as events in the present experience (Duncan et al. 2009). When they are using the describing method, construction personnel are likely to be focused on the present moment, sense their situation here and now and accurately describe the project and difficulties. It can then help them make fewer wrong decisions and meet clients' requirements, which can improve their personal satisfaction and project outcomes. Interpersonal performance is related to attachment to the community and attachment to other people in society. On the other hand, communication skills in the construction industry are also of paramount importance in keeping good relationships between colleagues, subordinates and bosses. Describing is useful for improving mutual communication among project team members, which helps them feel attached to the team. Hence, the more construction personnel use the describing method, the higher is their personal and interpersonal performance and the better are the project outcomes.

Awareness was negatively correlated with physical stress and emotional stress. In the face of threats, people normally face them (fight) or flee from them (flight; Cannon 1935). When construction personnel tackle problems with awareness, they can see stressful events as challenges rather than threats (and they can act with awareness, which induces optimism; Nes, Segerstrom and Sephton 2005). Their emotional reactions to stress may be minimised and their physical stress may be reduced because they perceive less threat. Using awareness coping behaviour, construction personnel can concentrate on their current work and their bodily sensations and thus they are not easily distracted. With awareness, construction personnel can pay attention to the potential risks in the site environment and improve their safety behaviours. They are also more likely to adjust their personal values to match the situation, which fosters a sense of belonging. However, awareness was also negatively correlated with intention to stay. One potential explanation is that once they are aware that the company is not suitable for their personal development, they may chose to leave.

Beginner's mind is a mindset that is willing to see everything as if for the first time. It guides us to be free of the expectations of our past experiences and it allows us to be receptive to new possibilities. It prevents us from getting stuck in the rut of our own expertise that leads personnel to complete tasks in ineffective ways (Henderson, Weeks and Hollingworth 1999). With this beginner's mind, construction personnel can pay attention to their colleagues and supervisors, communicate with less bias and negligence and have the capacity to arouse positive feelings in their colleagues. Hence, as a coping behaviour, beginner's mind is positively related to the interpersonal relationships of construction professionals.

6.8   Practical Implications

In sum, two types of problem-based coping behaviours (planful problem solving and instrumental support seeking) are positively related to four types of performance (i.e., personal satisfaction, interpersonal relationships, project outcomes and sense of belonging), while another two types of problem-based coping behaviours (confrontive coping and positive reappraisal) are only positively correlated with two types of performance (personal satisfaction and project outcomes). However, confrontive coping is also positively related to stress.

Thus, it is strongly recommended that construction personnel *use planful problem solving and instrumental support seeking to alter the stressful environment*. To help people plan out problem-solving procedures in advance and establish instrumental support, it is suggested that companies conduct partnering workshops (Chan et al. 2004) at the commencement of construction projects. It is also suggested that companies conduct value management workshops to solve problems as a team (Leung et al. 2013) and arrange occasional extra activities to develop team spirit (Gates 2003).

Consistent with previous studies, this study found that emotion-based coping behaviours are maladaptive. Three types of emotion-based coping behaviours (emotional discharge, escapism and self-controlling) can exacerbate the stress level of construction personnel. It is suggested that construction companies provide their employees with training for stress-coping strategies so that they can equip themselves with effective coping behaviours (Chan, Leung and Yu 2012). At the same time, it is recommended that construction organisations put effort into preventing their employees from using emotion-based coping behaviours (e.g., they can advertise the harm of emotion-based coping behaviours). In addition, construction companies can regularly organise formal and informal organisational gatherings, such as meals and ball games, in which construction personnel can relieve their stress and share their the successful experience in coping with stress with their colleagues.

This study finds that meaning-based coping behaviours help construction personnel reduce stress and improve their performance. Unlike problem- and emotion-based coping, meaning-based coping behaviour can be enhanced through formal and informal practices. Hence, it is recommended that construction personnel undertake mindfulness practices, including mindful stretching, body scanning, mindful sitting and mindful walking. These can be done in mindfulness-based stress reduction workshops and in daily practice. These exercises can help employees cultivate their inner side to handle any unexpected and complicated construction tasks in their daily work.

6.9   Summary

This chapter introduced and identified three groups of coping behaviours: problem-based (planful problem solving, positive reappraisal, confrontive coping and instrumental support seeking), emotion-based (emotional support seeking, emotional discharge, escapism and self-controlling) and meaning-based (MBSR attitudes). Based on the extensive literature in this area, a conceptual model of coping behaviours was developed.

This chapter also reported the results of questionnaire survey and case studies. Among the three groups of coping behaviours, construction personnel are most likely to use problem-based coping behaviours, especially planful problem solving, instrumental support seeking and positive reappraisal. Coping behaviours of construction personnel were compared in terms of age and gender. The results showed that young personnel used less problem-based (i.e., planful problem solving and instrumental support seeking) and meaning-based (i.e., present focus and awareness) coping behaviours, but more emotional coping behaviours (i.e., escapism) and beginner's mind. Male personnel tend to use confrontive coping, while female professionals prefer emotional support seeking.

This study also investigated the relationships between the coping behaviour, stress and performance of construction personnel. Four types of problem-based coping behaviours were positively related to most types of performance. Emotional support seeking, emotional discharge and escapism had no direct relationship with final performance, while all emotion-based coping behaviours (except emotional support seeking) were positively associated with stress. Present focus, non-judgmental and awareness were negatively related to stress, but non-judgmental was related to decreased performance. Describing and beginner's mind were positively related to performance. To cross-check the quantitative data, a case study was also conducted.

Based on both quantitative and qualitative data analyses, several practical implications were raised. It is recommended that construction personnel in stressful situations use problem-based and meaning-based coping behaviours, including planful problem solving, instrumental support seeking, present focus and awareness.

References

Aldwin, C.M. and Revenson, T. A. (1987) Does coping help? A re-examination of the relation between coping and mental health. *Journal of Personality and Social Psychology*, 53(2), 337.

Anderzen, I. and Arnetz, B.B. (1997) Psychophysiological reactions during the first year of a foreign assignment: Results of a controlled longitudinal study. *Work and Stress*, 11(4), 304–318.

Baer, R.A., Smith, G.T. and Allen, K.B. (2004) Assessment of mindfulness by self-report: The Kentucky Inventory of Mindfulness Skills. *Assessment*, 11(3), 191–206.

Baiden, B.K. and Price, A.D. (2011) The effect of integration on project delivery team effectiveness. *International Journal of Project Management*, 29(2), 129–136.

Bandura, A. (1992) Self efficacy mechanism in psychobiologic functioning. In R. Schwarzer (ed.), *Self Efficacy: Thought Control of Action*, 355–393. Washington, DC: Hemisphere.

Bays, J.C. (2009) *Mindful Eating: A Guide to Rediscovering a Healthy and Joyful Relationship with Food*. New York: Shambhala.

Billings, A.G. and Moos, R.H. (1984) Coping, stress and social resources among adults with unipolar depression. *Journal of Personality and Social Psychology*, 46(4), 877.

Bishop, S.R. (2002) What do we really know about mindfulness-based stress reduction? *Psychosomatic Medicine*, 64(1), 71–83.

Bishop, S.R., Lau, M., Shapiro, S., Carlson, L. Anderson, N.D., Carmody, J., . . . Devins, G. (2004) Mindfulness: A proposed operational definition. *Clinical psychology: Science and Practice*, 11(3), 230–241.

Blanchard-Fields, F., Chen, Y. and Norris, L. (1997) Everyday problem solving across the adult life span: Influence of domain specificity and cognitive appraisal. *Psychology and Aging*, 12(4), 684.

Blanchard-Fields, F., Mienaltowski, A. and Seay, R.B. (2007) Age differences in every-day problem-solving effectiveness: Older adults select more effective strategies for interpersonal problems. *Journal of Gerontology Series B: Psychological Sciences and Social Sciences*, 62(1), P61–P64.

Bowen, P., Edwards, P., Lingard, H. and Cattell, K. (2014) Workplace stress, stress effects and coping mechanisms in the construction industry. *Journal of Construction Engineering and Management*, 140(3).

Borker, D.R. (2012) Mindfulness practices for accounting and business education: A new perspective. *American Journal of Business Education*, 6(1), 41–56.

Brown, K.W. and Ryan, R.M. (2003) The benefits of being present: mindfulness and its role in psychological well-being. *Journal of Personality and Social Psychology*, 84(4), 822.

Buchheld, N., Grossman, P. and Walach, H. (2001) Measuring mindfulness in insight meditation (Vipassana) and meditation-based psychotherapy: The development of the Freiburg Mindfulness Inventory (FMI). *Journal for Meditation and Meditation Research*, 1(1), 11–34.

Cannon, W.B. (1935) Stresses and strains of homeostasis. *American Journal of Medical Science*, 189, 1–14.

Carmody, J. and Baer, R.A. (2008) Relationships between mindfulness practice and levels of mindfulness, medical and psychological symptoms and well-being in a mindfulness-based stress reduction program. *Journal of Behavioral Medicine*, 31(1), 23–33.

Carmody, J., Baer, R.A., Lykins, E. and Olendzki, N. (2009) An empirical study of the mechanisms of mindfulness in a mindfulness-based stress reduction program. *Journal of Clinical Psychology*, 65(6), 613–626.

Carter, S.M. and Chapman, S. (2003) Smoking, disease and obdurate denial: The Australian tobacco industry in the 1980s. *Tobacco Control*, 12, 23–30.

Carver, C.S., Scheier, M.F. and Weintraub, J.K. (1989) Assessing coping strategies: A theoretically based approach. *Journal of Personality and Social Psychology*, 56(2), 267.

Chan, Y.S.I. (2008) *Optimizing Stress and Performance of Hong Kong Construction Professionals: A Cultural Study*. Master of Philosophy, Department of Building and Construction, City University of Hong Kong.

Chan, Y.S.I. (2011) *Stress Management of Hong Kong Expatriate Construction Professionals in Mainland China*. Doctoral dissertation, Department of Building and Construction, City University of Hong Kong.

Chan Y.S. and Leung M.Y. (2013) The critical role of cultural values for coping behaviours of professionals in the stressful construction industry. *Engineering, Construction and Architectural Management*, 21(2).

Chan, I.Y.S., Leung, M.Y. and Yu, S.S.W. (2012) Managing the stress of Hong Kong expatriate construction professionals in Mainland China: A focus group study exploring individual coping strategies and organizational support. *Journal of Construction Engineering and Management*, 138(10), 1150–1160.

Chan, A. P., Chan, D. W., Chiang, Y. H., Tang, B. S., Chan, E. H., & Ho, K. S. (2004). Exploring critical success factors for partnering in construction projects. *Journal of Construction Engineering and Management*, 130(2), 188-198.

Chang, E.M., Daly, J.W., Hancock, K.M., Bidewell, J., Johnson, A., Lambert, V.A. and Lambert, C.E. (2006) The relationships among workplace stressors, coping methods, demographic characteristics and health in Australian nurses. *Journal of Professional Nursing*, 22(1), 30–38.

Christman, N.J., McConnell, E.A., Pfeiffer, C., Webster, K.K., Schmitt, M. and Ries, J. (1988) Uncertainty, coping and distress following myocardial infarction: Transition from hospital to home. *Research in Nursing and Health*, 11(2), 71–82.

Chung, M.C., Symons, C., Gilliam, J. and Kaminski, E.R. (2010). Stress, psychiatric co-morbidity and coping in patients with chronic idiopathic urticaria. *Psychology and Health*, 25(4), 477–490.

Cohen, S., Kessler, R.C. and Gordon, L.U. (1995) Strategies for measuring stress in studies of psychiatric and physical disorders. *Measuring Stress: A Guide for Health and Social Scientists*. Oxford: Oxford University Press.

Crane, R. (2009) *Mindfulness-based Cognitive Therapy: Distinctive Features*. New York: Routledge.

de Ribaupierre, A. (2000) Working memory and attentional control. In W.J. Perrig and A. Grob (eds), *Control of Human Behavior, Mental Processes and Consciousness: Essays in Honor of the 60th Birthday of August Flammer*. Mahwah, NJ: Lawrence Erlbaum.

de Rijk, A.E., Le Blanc, P., Shaufeli, W. and de Jonge, J. (1998) Active coping and need for control as moderators of the job–demand–control model: effects on burnout. *Journal of Occupational and Organizational Psychology*, 71, 1–18.

Djebarni, R. (1996) The impact of stress in site management effectiveness. *Construction Management and Economics*, 14(4), 281–293.

Duncan, L. G., Coatsworth, J. D., & Greenberg, M. T. (2009). Pilot study to gauge acceptability of a mindfulness-based, family-focused preventive intervention. *The journal of primary prevention*, 30(5), 605-618.

Dyson, R. and Renk, K. (2006) Freshmen adaptation to university life: Depressive symptoms, stress and coping. *Journal of Clinical Psychology*, 62(10), 1231–1244.

Esteve, L.G., Valdés, M., Riesco, N., Inmaculada, J. and de Flores, T. (1992) Denial mechanisms in myocardial infarction: Their relations with psychological variables and short-term outcome. *Journal of Psychosomatic Research*, 36(5), 491–496.

Farhall, J. and Gehrke, M. (2011) Coping with hallucinations: Exploring stress and coping framework. *British Journal of Clinical Psychology*, 36(2), 259–261.

Fincham, F.D., Beach, S.R. and Davila, J. (2004). Forgiveness and conflict resolution in marriage. *Journal of Family Psychology*, 18, 72–81.

Folkman, S. (1984) Personal control and stress and coping processes: A theoretical analysis. *Journal of Personality and Social Psychology*, 46(4), 839–852.

Folkman, S. (1987) Positive psychological states and coping with severe stress. *Social Science and Medicine*, 45, 1207–1221.

Folkman, S. (2010) Stress, coping and hope. *Psycho-Oncology*, 19, 901–908.

Folkman, S. and Lazarus, R.S. (1988) *Manual for the Ways of Coping Questionnaire*. Palo Alto, CA: Consulting Psychologists Press.

Folkman, S., Aldwin, C. and Lazarus, R. (1981) *The relationship between locus of control, cognitive appraisal and coping*. Paper presented at the meetings of the American Psychological Association. Los Angeles, CA.

Folkman, S., Lazarus, R.S., Dunkel-Schetter, C., DeLongis, A. and Gruen, R.J. (1986) Dynamics of a stressful encounter: Cognitive appraisal, coping and encounter outcomes. *Journal of Personality and Social Psychology*, 50(5), 992.

Folkman, S., Lazarus, R.S., Pimley, S. and Novacek, J. (1987) Age differences in stress and coping processes. *Psychology and Aging*, 2(2), 171.

Folkman, S. and Moskowitz, J.T. (2000) Positive affect and the other side of coping. *American Psychologist*, 55(6), 647–654.

Folkman, S., Schaefer, C. and Lazarus, R.S. (1979) Cognitive processes as a mediator of stress and coping. In V. Hamilton and D.M. Warburton (eds), *Human Stress and Cognition: An Information Processing Approach*. Chichester: Wiley.

Francis, V. and Lingard, H. (2004) A quantitative study of work-life experiences in the public and private sectors of the Australian construction industry. Construction Industry Institute Australia, *Brisbane*, 142.

Franco, C., Mañas, I., Cangas, A.J., Moreno, E. and Gallego, J. (2010) Reducing teachers' psychological distress through a mindfulness training program. *Spanish Journal of Psychology*, 13(2), 655–666.

Frese, M. (1986) Coping as a moderator and mediator between stress at work and psychosomatic complaints. *Dynamics of Stress: Physiological, Psychological and Social Perspectives*, 183–206.

French, H. C. (2004). Occupational stresses and coping mechanisms of therapy radiographers-a qualitative approach. *Journal of Radiotherapy in Practice*, 4(01), 13-24.

Gates, E. (2003) Workplace stress counselling, *Occupational Safety and Health*, 33(5), 40-44.

Germer, C.K., Siegel, R.D. and Fulton, P.R. (eds) (2005) *Mindfulness and Psychotherapy*. New York: Guilford Press.

Glidden, L.M., Billings, F.J. and Jobe, B.M. (2006) Personality, coping style and well-being of parents rearing children with developmental disabilities. *Journal of Intellectual Disability Research*, 50(12), 949–962.

Greenglass, E. (1997) The impact of social support on the development of burnout in teachers: Examination of a model. *Work and Stress*, 11(3), 267–278.

Greenglass, E., Schwarzer, R. and Taubert, S. (1999) The proactive coping inventory (PCI): A multidimensional research instrument. In *Proceedings of the 20th International Conference of the Stress and Anxiety Research Society (STAR)*, Cracow, Poland.

Grossman, P., Niemann, L., Schmidt, S. and Walach, H. (2004) Mindfulness-based stress reduction and health benefits – A meta-analysis. *Journal of Psychosomatic Research*, 57(1), 35–44.

Gunning, J.G. and Keaveney, M. (1998) A transverse examination of occupational stress among a cross-disciplinary population of Irish construction professionals. *Management*, 1, 98–106.

Hair, J.F.J., Anderson, R.E., Tatham, R.L. and Black, W.C. (1998) *Multivariate Data Analysis*. New Jersey: Prentice Hall.

Hanh, T.N. (1998) *The nature of self*. Retrieved from http://mountainsangha.org/the-nature-of-self/

Havik, O.E. and Maeland, J.G. (1988) Verbal denial and outcome in myocardial infarction patients. *Journal of Psychosomatic Research*, 32(2), 145–157.

Hawley, L., Schwartz, D., Bieling, P., Irving, J., Corcoran, K. and Farb, N. (2014) Mindfulness practice, rumination and clinical outcome in mindfulness-based treatment. *Cognitive Therapy and Research*, 38(1), 1–9.

Hayes, S.C., Strosahl, K.D. and Wilson, K.G. (1999) *Acceptance and Commitment Therapy: An Experiential Approach to Behavior Change*. New York: Guilford Press.

Haynes, N.S. and Love, P.E. (2004) Psychological adjustment and coping among construction project managers. *Construction Management and Economics*, 22(2), 129–140.

Heaversedge, J. (2012) *The Mindful Manifesto: How Doing Less and Noticing More Can Treat Illness, Relieve Stress and Help Us Cope with the 21st Century*. London: Hay House.

Henderson, J.M., Weeks, P.A., Jr and Hollingworth, A. (1999). The effects of semantic consistency on eye movements during complex scene viewing. *Journal of Experimental Psychology: Human Perception and Performance*, 25, 210–228.

Hoedaya, D. and Anshel, M.H. (2003) Use and effectiveness of coping with stress in sport among Australian and Indonesian athletes. *Australian Journal of Psychology*, 55(3), 159–165.

Holahan, C.J., Moos, R.H., Holahan, C.K. and Brennan, P.L. (1997) Social context, coping strategies and depressive symptoms: An expanded model with cardiac patients. *Journal of Personality and Social Psychology*, 72(4), 918.

Holmes, D.A. (2007) *Psyche's Palace: How the Brain Generates the Light of the Soul*. New York: Library of Consciousness.

Holroyd, K.A. and Lazarus, R.S. (1982) Stress, coping and somatic adaptation. *Handbook of Stress: Theoretical and Clinical Aspects*. New York: Free Press.

Howard, R. and McKillen, M. (1990) Extraversion and performance in the perceptual maze test. *Personality and Individual Differences*, 11(4), 391–396.

Israel, B. A., Schurman, S.J. and House, J.S. (1989) Action research on occupational stress: Involving workers as researchers. *International Journal of Health Services*, 19(1), 135–155.

Jain, S., Shapiro, S.L., Swanick, S., Roesch, S.C., Mills, P.J., Bell, I. and Schwartz, G.E. (2007) A randomized controlled trial of mindfulness meditation versus relaxation training: Effects on distress, positive states of mind, rumination and distraction. *Annals of Behavioral Medicine*, 33(1), 11–21.

Kabat-Zinn, J. (1982) An outpatient program in behavioral medicine for chronic pain patients based on the practice of mindfulness meditation: Theoretical considerations and preliminary results. *General Hospital Psychiatry*, 4(1), 33–47.

Kabat-Zinn, J. (1990) *Full Catastrophe Living: Using the Wisdom of Your Mind and Body to Face Stress, Pain and Illness*. New York: Delacorte.

Kabat-Zinn, J. (2003) Mindful yoga: Movement and meditation. *Yoga International*, 70, 86–93.

Kabat-Zinn, J., Massion, A.O., Kristeller, J., Peterson, L.G., Fletcher, K.E. and Pbert, L. (1992) Effectiveness of a meditation-based stress reduction program in the treatment of anxiety disorders. *American Journal of Psychiatry*, 149, 936–943.

Karimzade, A. and Besharat, M.A. (2011) An investigation of the relationship between personality dimensions and stress coping styles. *Procedia-Social and Behavioral Sciences*, 30, 797–802.

Keavney, G. and Sinclair, K.E. (1978) Teacher concerns and teacher anxiety: A neglected topic of classroom research. *Review of Educational Research*, 48, 273–290.

Kenrick, D.T., Neuberg, S.L. and Cialdini, R.B. (2010) *Social Psychology: Goals in Interaction*. Boston: Allyn & Bacon.

King, K.B. (1985) Measurement of strategies, concerns and emotional response in patients undergoing coronary artery bypass grafting. *Heart and Lung*, 14, 579–586.

Kraimer, M.L., Wayne, S.J. and Jaworski, R.A. (2001) Sources of support and expatriate performance: The mediating role of expatriate adjustment. *Personnel Psychology*. 54(1), 71–79.

Krasner, M.S., Epstein, R.M., Beckman, H., Suchman, A.L., Chapman, B., Mooney, C.J. and Quill, T.E. (2009) Association of an educational program in mindful communication with burnout, empathy and attitudes among primary care physicians. *Journal of the American Medical Association*, 302(12), 1284–1293.

Kristeller, J.L. and Hallett, C.B. (1999) An exploratory study of a meditation-based intervention for binge eating disorder. *Journal of Health Psychology*, 4(3), 357–363.

Krohne, H.W. (1986) *Coping with Stress: Dispositions, Strategies and the Problem of Measurement*. New York: Plenum.

Lam, A.G. and Zane, N.W. (2004) Ethnic differences in coping with interpersonal stressors: A test of self-construals as cultural mediators. *Journal of Cross-Cultural Psychology*, 35(4), 446–459.

Lambert, V.A. and Lambert, C.E. (2008) Nurses' workplace stressors and coping strategies. *Indian Journal of Palliative Care*, 14(1), 38.

Laranjeira, C.A. (2012) The effects of perceived stress and ways of coping in a sample of Portuguese health workers, *Journal of Clinical Nursing*, 21(11–12), 1755–1762.

Latack, J.C. (1986) Coping with job stress: Measures and future directions for scale development. *Journal of Applied Psychology*, 71(3), 377.

Latack, J.C. and Havlovic, S.J. (1992) Coping with job stress: A conceptual evaluation framework for coping measures. *Journal of Organizational Behavior*, 479–508.

Laux, L. (1986) A self-presentational view of coping and stress. In M.H. Appley and R. Trumbull (eds), *Dynamics of Stress: Physiological, Psychological and Social Perspectives*. New York: Plenum.

Lazarus, R.S. and Folkman, S. (1984) *Stress, Appraisal and Coping*. New York: Springer.

Lefrancois, G.R. (1996) *The Lifespan*. Belmont, CA: Wadsworth.

Leiter, M.P. (1991) Coping patterns as predictors of burnout: The function of control and escapist coping patterns. *Journal of Organizational Behavior*, 12(2), 123–144.

Lennerlöf, L. (1988) Learned helplessness at work. *International Journal of Health Services*, 18(2), 207–222.

Leung, M.Y., Liu, A.M. and Wong, M.M.K. (2006) Impact of stress-coping behaviour on estimation performance. *Construction Management and Economics*, 24(1), 55–67.

Leung, M. Y., Yu, J., & Liang, Q. (2013). Analysis of the Relationships between Value Management Techniques, Conflict Management, and Workshop Satisfaction of Construction Participants. *Journal of Management in Engineering*, 30(3).

Leung, M.Y., Wong, M.K. and Oloke, D. (2003) Coping behaviors of construction estimators in stress management. *Proceedings of 19th Annual Conference of the Association of Researchers in Construction Management (ARCOM 2003)*, UK.

Lingard, H. (2003) The impact of individual and job characteristics on 'burnout' among civil engineers in Australia and the implications for employee turnover. *Construction Management and Economics*, 21(1), 69–80.

Livneh, H. (1999) Psychosocial adaptation to heart diseases: The role of coping strategies. *Journal of Rehabilitation*, 11, 401–418.

Lösel, F. and Bliesener, T. (1990) Resilience in adolescence: A study on the generalizability of protective factors. In K. Hurrelmann and F. Lösel (eds), *Health Hazards in Adolescence*, 299–320. Berlin: De Gruyter.

Lowery, B.J., Jacobsen, B.S., Cera, M.A., McIndoe, D., Kleman, M. and Menapace, F. (1992) Attention versus avoidance: Attributional search and denial after myocardial infarction. *Heart and Lung*, 21(6), 523.

Madu, S.N. and Roos, J.J. (2006) Depression among mothers with preterm infants and their stress-coping strategies. *Social Behavior and Personality*, 34(7), 877–890.

Mathieu, J. (2009) What should you know about mindful and intuitive eating? *Journal of the American Dietetic Association*, 109(12), 1982–1982.

McDonald, I.H. and Stocks, J.G. (1965) Prolonged nasotracheal intubation: A review of its development in a paediatric hospital. *British Journal of Anaesthesia*, 37(3), 161–173.

Moore, A. (2008) *Meditation*. New York: Rosen.

Morone, N,E., Greco, C.M. and Weiner, D.K. (2008) Mindfulness meditation for the treatment of chronic low back pain in older adults: A randomized controlled pilot study. *Pain*, 134, 310–319.

Murphy, L.R. (1988) Workplace interventions for stress reduction and prevention. In C.L. Cooper and R. Payne (eds), *Causes, Coping and Consequences of Stress at Work*. New York: Wiley.

Nes, L.S., Segerstrom, S.C. and Sephton, S.E. (2005) Engagement and arousal: Optimism's effects during a brief stressor. *Personality and Social Psychology Bulletin*, 31(1), 111–120.

Newton, T.J. and Keenan, A. (1985) Coping with work-related stress. *Human Relations*, 38(2), 107–126.

Oakland, S. and Ostell, A. (1996) Measuring coping: A review and critique. *Human Relations*, 49(2), 133–155.

O'Connor, D.B. and Shimizu, M. (2002) Sense of personal control, stress and coping style: A cross-cultural study. *Stress and Health*, 18(4), 173–183.

Olpin, M. and Hesson, M. (2012) *Stress Management for Life: A Research-Based Experiential Approach*. New York: Wadsworth.

Ott, M.J., Longobucco-Hynes, S. and Hynes, V.A. (2002) Mindfulness meditation in pediatric clinical practice. *Pediatric nursing*, 28(5), 487–495.

Owasoyo, J.O., Neri, D.F. and Lamberth, J.G. (1992) Tyrosine and its potential use as a countermeasure to performance decrement in military sustained operations. *Aviation, Space and Environmental Medicine*, 63(5), 364–369.

Parker, J.D. and Endler, N.S. (1996) Coping and defense: A historical overview. In M. Zeidner and N. Endler (eds.), *Handbook of Coping: Theory, Research, Applications*. New York: Wiley.

Pearlin, L.I. and Schooler, C. (1978) The structure of coping. *Journal of Health and Social Behavior*, 2–21.

Penley, J.A., Tomaka, J. and Wiebe, J.S. (2002) The association of coping to physical and psychological health outcomes: A meta-analytic review. *Journal of Behavioral Medicine*, 25(6), 551–603.

Poon, W.C., Lee, C.K.C. and Ong, T.P. (2012) Undergraduates' perception on causes, coping and outcomes of academic stress: Its foresight implications to university administration. *International Journal of Foresight and Innovation Policy*, 8(4), 379–403.

Quinn, T., Bussell, J.L. and Heller, B. (2010) *Fully Fertile: A Holistic 12-Week Plan for Optimal Fertility*. London: Findhorn Press.

Rehm, J., Room, R., Graham, K., Monteiro, M., Gmel, G. and Sempos, C.T. (2003) The relationship of average volume of alcohol consumption and patterns of drinking to burden of disease: An overview. *Addiction*, 98(9), 1209–1228.

Richards, D.A. and Schat, A.C. (2011) Attachment at (not to) work: Applying attachment theory to explain individual behavior in organizations. *Journal of Applied Psychology*, 96(1), 169.

Rivera, R. (2008) The work adjustment process of expatriate managers: An exploratory study in Latin America. *Journal of Centrum Cathedra*, 1(1).

Robins, C.J. (2003) Zen principles and mindfulness practice in dialectical behavior therapy. *Cognitive and Behavioral Practice*, 9(1), 50–57.

Roemer, L. and Orsillo, S.M. (2002) Expanding our conceptualization of and treatment for generalized anxiety disorder: Integrating mindfulness/acceptance-based approaches with existing cognitive-behavioral models. *Clinical Psychology*, 9(1), 54–68.

Sallis, J.F., Grossman, R.M., Pinski, R.B., Patterson, T.L. and Nader, P.R. (1987) The development of scales to measure social support for diet and exercise behaviors. *Preventive Medicine*, 16(6), 825–836.

Salmon, P., Lush, E., Jablonski, M. andSephton, S.E. (2009) Yoga and mindfulness: Clinical aspects of an ancient mind/body practice. *Cognitive and Behavioral Practice*, 16(1), 59–72.

Schafer, W. (2000) *Stress Management for Wellness*. New York: Wadsworth.

Schwarzer, R. (1999) Self-regulatory processes in the adoption and maintenance of health behaviors. *Journal of Health Psychology*, 4(2), 115–127.

See, C.M. and Essau, C.A. (2010) Coping strategies in cross-cultural comparison. *Psychologie–Kultur–Gesellschaft*, 161–173.

Segal, Z.V, Williams, J.M.G. and Teasdale, J.D. (2002) *Mindfulness-Based Cognitive Therapy for Depression: A New Approach to Preventing Relapse*. New York: Guilford Press.

Segal, Z.V., Williams, J.M.G. and Teasdale, J.D. (2012) *Mindfulness-Based Cognitive Therapy for Depression*. New York: Guilford Press.

Spector, P.E. (1986) Perceived control by employees: A meta-analysis of studies concerning autonomy and participation at work. *Human Relations*, 39(11), 1005–1016.

Stahl, B. and Goldstein, E. (2010) *A Mindfulness-Based Stress Reduction Workbook*. New York: New Harbinger.

Stein, N., Trabasso, T., Folkman, S. and Richards, A.T. (1997) Appraisal and goal processes as predictors of psychological well-being in bereaved caregivers. *Journal of Personality and Social Psychology*, 72, 872–884.

Stewart, M.J., Hirth, A.M., Klassen, G., Makrides, L. and Wolf, H. (1997) Stress, coping and social support as psychosocial factors in readmissions for ischaemic heart disease. *International Journal of Nursing Studies*, 34(2), 151–163.

Stokes, J.P. and Wilson, D.G. (1984) The inventory of socially supportive behaviours: Dimensionality, prediction and gender differences. *American Journal of Community Psychology*, 12, 53–69.

Strough, J., McFall, J.P., Flinn, J.A. and Schuller, K.L. (2008) Collaborative everyday problem solving among same-gender friends in early and later adulthood. *Psychology and Aging*, 23(3), 517.

Sullivan, A. (2002) Gender differences in coping strategies of parents of children with Down syndrome. *Down Syndrome Research and Practice*, 8(2), 67–73.

Tanskanen, K., Holmström, J., Elfving, J. and Talvitie, U. (2008) Vendor-managed-inventory (VMI) in construction. *International Journal of Productivity and Performance Management*, 58(1), 29–40.

Terry, D.J. (2011) Stress, coping and coping resources as correlates of adaptation in myocardial infarction patients. *British Journal of Clinical Psychology*, 31(2), 215–225.

Thoits, P.A. (1995) Stress, coping and social support processes: Where are we? What next? *Journal of Health and Social Behavior*, 53–79.

Tweed, R.G., White, K. and Lehman, D.R. (2004) Culture, stress and coping internally- and externally-targeted control strategies of European Canadians, East Asian Canadians and Japanese. *Journal of Cross-Cultural Psychology*, 35(6), 652–668.

Varhol, P. (2000) Identify and manage work-related stress. *Electronic Design*, 48(26), 123–124.

Vingerhoets, A.J. and Van Heck, G.L. (1990) Gender, coping and psychosomatic symptoms. *Psychological Medicine*, 20(1), 125–135.

Wang, X. and Nayir, D.Z. (2006) How and when is social networking important? Comparing European expatriate adjustment in China and Turkey. *Journal of International Management*, 12(4), 449–472.

Violanti, J.M. (1993) What does high stress police training teach recruits? An analysis of coping. *Journal of Criminal Justice*, 21(4), 411–417.

Watson, T.L. and Blanchard-Fields, F. (1998) Thinking with your head and your heart: Age differences in everyday problem-solving strategy preferences. *Aging, Neuropsychology and Cognition*, 5(3), 225–240.

Watts, J.H. (2007) Allowed into a man's world' meanings of work–life balance: Perspectives of women civil engineers as 'minority' workers in construction. *Gender, Work and Organization*, 16(1), 37–57.

Webster, K.K. and Christman, N.J. (1988) Perceived uncertainty and coping post myocardial infarction. *Western Journal of Nursing Research*, 10(4), 384–400.

Whittington, R. and Wykes, T. (1994) Going in strong: Confrontive coping by staff. *Journal of Forensic Psychiatry*, 5(3), 609–614.

Wills, T.A. (1991) Social support and interpersonal relationships. *Prosocial Behavior*, 12, 265.

Xianyu, Y. and Lambert, V.A. (2006) Investigation of the relationships among workplace stressors, ways of coping and the mental health of Chinese head nurses. *Nursing and Health Sciences*, 8(3), 147–155.

Yip, B. and Rowlinson, S. (2006) Coping strategies among construction professionals: Cognitive and behavioural efforts to manage job stressors. *Journal for Education in the Built Environment*, Vol. 1(2), 70–79.

Conclusions

This chapter firstly presents an integrated stress management model for construction personnel and then summarises the results of previous research studies. Based on these findings, both practical and academic recommendations are made in order to optimise the performance of individuals, organisations and the industry as a whole.

7.1 Stress Management for Construction Personnel

Stress theories can be traced back to the early twentieth century, when stress began to be considered as a form of general anxiety rooted in the cares of life (Cannon 1914). Stress has been widely studied from the perspectives of psychobiology (Langeland and Olff 2008), sociology (Pearlin 1989) and psychiatry (Puca et al. 1999). Various theories have been developed, from arousal theories (such as the fight or flight responses, eustress versus distress and the Yerkes–Dodson law: Cannon 1927; Selye 1964; Yerkes and Dodson 1908) to the appraisal and regulatory theories (transactional stress, controlled and compensatory processing and stress-adaptation: Hancock and Warm 1989; Hockey 1997; Lazarus and Folkman 1984; Parasuraman and Rizzo 2008). However, few comprehensive models have been proposed, especially for construction personnel. The authors have therefore conducted a series of scientific studies to investigate the complicated relationships between stressors (stimuli), stress, coping behaviours and performance for this occupational group.

Based on the various stress theories (Chapter 2) and the results of research studies (Chapters 3–6), we propose an integrated stress management model illustrating the interrelationships of various stressors, stress, performance and coping behaviours for construction personnel (see Figure 7.1).

Stress Management in the Construction Industry, First Edition.
Mei-yung Leung, Isabelle Yee Shan Chan and Cary L. Cooper.
© 2015 John Wiley & Sons, Ltd. Published 2015 by John Wiley & Sons, Ltd.

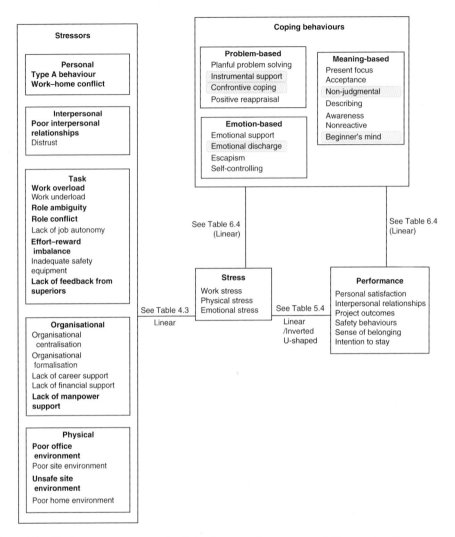

Figure 7.1   Final stressor–stress–coping behaviours–performance model for construction personnel. Note: Highlighted items refer to coping behaviours related to both stress and performance; stressors in **bold** have significant relationships with all three kinds of stress.

7.1.1   Stress of Construction Personnel

The extensive literature review presented in Chapter 3 delineated three kinds of stress, namely work, physical and emotional. *Work stress* arises from the deviation between the expected and actual ability of an individual to handle a job task (Cox 1993; Leung et al. 2005, Leung, Liu and Wong 2006; Leung, Sham and Chan 2007; Leung, Chan and Olomolaiye 2008). *Physical stress* mainly focuses on physical symptoms (such as headaches, appetite loss, sleep disorders, musculoskeletal pain, high blood pressure and excessive sweating) (Chrousos and Gold 1992; Leung, Chan and Dongyu 2011; Teasdale and McKeown 1994). *Emotional stress* results in symptoms such as worry, fatigue, feelings of being used up and emotionally drained, frustration, unhappiness, loss of temper and anxiety (Daniels and Guppy 1994; Leung, Chan and Yuen 2010; Nyssen et al. 2003). For construction personnel, the main indicator of work stress was the number of projects in which they are involved. Otherwise, they mainly suffered from appetite loss and musculoskeletal pain (physical stress) and worry, fatigue, feelings of being used up and anxiety (emotional stress).

The stress levels of different groups of construction personnel in terms of professional discipline, age and gender were also compared. *Quantity surveyors and construction workers faced more physical stress than project managers.* Due to the physical effort required of construction workers, they often suffered from physical symptoms such as headaches, sweating and back pain. Quantity surveyors must often handle large amounts of paperwork and repetitive tasks, which may easily cause appetite loss. *The work stress of construction personnel aged 50 or above was lower than for other age groups.* Compared with younger personnel, with limited experience and ability, those aged 50 or above will have amassed adequate experience and knowledge and hence be more capable of handling their work tasks. In terms of gender, as females tend to be more aware of their emotional states, female personnel were found to suffer from a higher level of emotional stress than their male counterparts.

At the same time, the results also indicated *significant interrelationships between work, physical and emotional stress.* A mismatch of actual and expected ability to carry out the work induces high levels of work stress, which leads to physical and emotional problems. On the other hand, construction personnel with high levels of physical and emotional stress found this influenced their job performance and reduced their ability to cope and so, in turn, caused them work stress.

7.1.2   Stressors for Construction Personnel

A stressor is a stimulus that has the potential to elicit a stress reaction (Ganster and Rosen 2013; Selye 1976). Stressors for construction personnel can generally be divided into five major categories: personal, interpersonal, task, organisational and physical (see Chapter 4). Across these five categories, 21 stressors were identified, including personal (e.g. Type A behaviour and work–home conflict: Caplan and Jones 1975; Emslie, Hunt and Macintyre 2004; Leung, Sham and Chan 2007); interpersonal (e.g. poor relationships and distrust: Cooper 2001; Djebarni and Lansley 1995; Leung et al. 2005); task (e.g. work overload, work underload, role ambiguity, lack of job autonomy, effort–reward imbalance, inadequate safety equipment and lack of feedback from superiors: Djebarni 1996; Leung et al. 2005; Leung, Sham and Chan 2007; Yip 2008); organisational (e.g. organisational centralisation, organisational formalisation and lack of career support, lack of financial support and lack of manpower support: Eisenberger et al. 1986; Francesco and Gold 2005; Lazarus and Folkman 1984); and physical stressors (e.g. poor office environment, poor site environment, unsafe site environment and poor home environment: Driskell and Salas 1991; Leung, Chan, Chong and Sham 2008; Leung, Chan and Yu 2012; Quick and Quick 1997). Among these stressors, inadequate safety equipment, poor site environment and organisational formalisation were identified as the most significant by the construction personnel in these studies.

The results of the comparison of stressors indicated no significant differences in terms of professional discipline or gender. However, significant differences were found across age groups. Construction personnel aged 30–39 reported significantly higher levels of distrust of their colleagues and fellow team members than those aged 40–49. *Younger construction personnel are more likely to encounter work overload, work underload, lack of job autonomy, effort–reward imbalance and a poor office or site environment.* Younger personnel, as "freshmen" in the construction industry, normally have limited knowledge and experience to handle daily work and must follow others' instructions to complete their tasks (Keenan and Newton 1987). They are often assigned simple and repetitive tasks with little autonomy. Even though their work may be simple, less experienced personnel may still need to invest additional effort to complete those tasks and also often work in a poor environment. On the other hand, construction personnel aged 50 or above already earn more and may have little appetite for promotion compared with other age groups. Therefore, in general, they are less susceptible to stressors like lack of autonomy, poor physical environment and work underload.

In order to investigate the interrelationships between stressors and their impact on the three kinds of stress (work, physical and emotional), a Pearson correlation analysis was conducted. Personal (Type A behaviour and work–home conflict), interpersonal (poor interpersonal relationships), task (work overload, role ambiguity, role conflict, effort–reward imbalance and lack of feedback from superiors), organisational (lack of manpower support) and environmental stressors (poor office environment and unsafe site environments) were all highly correlated with forms of stress.

Both *Type A behaviour and work–home conflict* were positively related to all three kinds of stress. Construction personnel with Type A behaviours are competitive and achievement-oriented and easily lose their temper. They reported often completing their tasks despite a high level of stress, which induced physical and emotional problems. Work–home conflict not only influences the daily lives of construction personnel, but also affects their ability to perform their work and the amount of effort they invest. This further induces work stress and causes physical and emotional symptoms such as headaches, anxiety and worry (Greenhaus and Parasuraman 1987).

In terms of interpersonal stressors, *poor interpersonal relationships* was positively related to three kinds of stress, while distrust only affected emotional stress. Poor interpersonal relationships often cause emotional problems for construction personnel and also reduce their working ability and physical health (Beehr and Jex 2001). *Distrust* among construction personnel can easily cause worries and depression, which can also result in emotional stress.

Looking at task stressors, work overload, work underload, role ambiguity, role conflict, effort–reward imbalance, inadequate safety equipment and lack of feedback from superiors were all correlated with stress. *Work overload* can cause work stress if the tasks required exceed the individual's ability to perform them. It can also lead to prolonged overtime, which can undermine physical and emotional well-being. The results also indicated that *work underload* was positively related to physical stress. Work underload may result in boredom, leading to physical problems such as appetite loss and sleep disorders (Leung, Chan and Olomolaiye, 2008; Loosemore and Waters 2004). *Role ambiguity, role conflict and effort–reward imbalance* were also interrelated. The mismatch between responsibility, effort and reward may cause frustration, reduce ability and ultimately stimulate work, physical or emotional stress. Inadequate safety equipment was also positively related to emotional and physical stress. Safety equipment is used to prevent accidents or injuries to construction professionals and workers, which is beneficial to their physical health (e.g. they may wear ear plugs to prevent hearing damage due to excessive noise). If safety equipment is adequate, workers may feel safer and less anxious, leading to a lower level of emotional stress. Feedback from superiors is also essential to construction personnel, because it indicates satisfaction with their performance, which is closely related to reward and career development. A lack of feedback from superiors can easily cause someone to worry about their future, leading to physical and work stress.

In terms of organisational stressors, organisational centralisation and lack of career support were positively associated with emotional stress, while lack of manpower support was related to all three kinds. In a *centralised organisational structure*, employees are subject to a hierarchy of authority and have limited involvement in decisions (Andrews et al. 2009). They have limited job authority, which in turn predicts depression, low self-esteem and job dissatisfaction (Beehr 1995; Leung, Sham and Chan 2007), resulting in emotional stress. *Lack of career support* limits the career development of construction personnel and causes frustration and worry about their future. *Lack of manpower support* influences relationships between colleagues, which can also subsequently arouse serious work, physical and emotional stress.

All the physical stressors (i.e., poor office, site and/home environment and unsafe site environment) were positively associated with stress. Construction personnel who experienced *poor office and home environments* tended to suffer from work stress. Moreover, *poor office and site environments* (in terms of noise, air quality and temperature) had a further impact on physical and emotional health. Working on an *unsafe site* means that workers need to keep an eye on potential risks, which distracts them from tasks and induces work stress (Leung, Chan and Yu 2012). Incidents often happen in such an environment, which not only directly affects physical health but can also lead to serious emotional difficulties.

7.1.3   The Consequences of Stress for Construction Personnel

The consequences of stress for the performance of construction personnel were discussed in Chapter 5. Excessive stress leads to poorer performance, while too little stress can cause boredom and disinterest (Gmelch

1982; Selye 1976; Zaccaro and Riley 1987). Hence, stress levels that are either too high or too low can lead to impaired performance, while moderate levels of stress can result in optimum delivery (Yerkes and Dodson 1908).

The consequences of stress can be categorised into three aspects: *personal (including personal satisfaction and interpersonal relationships), task (including project outcomes and safety behaviours) and organisational (sense of belonging and intention to stay) performance* (Black and Gregersen 1990; Leung et al. 2005, Leung, Liu and Wong 2006; Leung et al. 2008; Leung, Chan and Yuen 2010). Out of the six types, interpersonal relationships achieved the highest mean score and safety behaviours the lowest. Construction personnel emphasised the need for collaboration among various parties for the implementation of projects. Hence, they considered relationships to be the most important performance indicator. On the other hand, given their propensity to overemphasise productivity, construction workers may be more likely to try to focus on getting the job done at the expense of safety (Mullen 2004). Safety performance therefore received a relatively low score from this group.

In order to compare the performances of construction personnel in terms of age and gender, a one-way between-group analysis of variance (ANOVA) was conducted. The results showed that *younger construction personnel gave significantly lower scores to interpersonal relationships and intention to stay.* Because they have just started their careers, younger construction personnel generally lack the skills and experience to handle more complicated problems. It is inevitable that they will make mistakes. Hence, they are more likely to be blamed by senior colleagues, affecting their interpersonal relationships. In addition, younger personnel might encounter difficulties in adapting to their jobs and to the construction industry in general. They may consequently report lower intention to stay in their first few years in the organisation (Reichers, Wanous and Steele 1994). Their relative lack of competence may be one of the factors contributing to this.

Male construction personnel gave project outcomes significantly higher scores than female. In the male-dominated construction industry, men are more likely to be assigned critical tasks and given more responsibility. Due to their different social roles, men are also often more able to invest greater time and effort in their work than women, who may also be burdened by domestic responsibilities. This may explain why men gave higher scores to this performance factor.

The relationships between stress and the different types of performance were also investigated using a Pearson correlation analysis. The results demonstrated significant inverted U-shaped relationships between work stress and two types of performance (personal satisfaction and safety behaviour); emotional stress and four types (interpersonal relationships, project outcomes, sense of belonging and intention to stay); and physical stress and three types (personal satisfaction, interpersonal relationships and intention to stay). Work stress was also negatively related to personal satisfaction, interpersonal relationships and project outcomes. Too high a level of *work stress* leads to reduced productivity, as well as health problems such as depression, anxiety and physical pain, all of which impair performance and therefore affect personal satisfaction. Moreover, chronic exposure to high levels of stress may result in construction personnel ignoring safety requirements, leading to accidents (Health and Safety Executive 2006). On the other hand, too little work stress results in boredom and lack of motivation, which also leads to job dissatisfaction and unsafe behaviours. Hence, for construction personnel, neither too much nor too little work stress is desirable. A moderate level will lead to optimum safety performance.

Construction personnel who are suffering from *physical stress* symptoms like headaches and loss of appetite may find it difficult to concentrate on their jobs. Working under physical stress for prolonged periods often reduces productivity, influences communication and collaboration with others, decreases job satisfaction and ultimately reduces the intention to stay with an organisation (Leung, Chan and Dongyu 2011). However, the inverted U-shaped relationship indicates that too little physical stress is also associated with poor performance. Given that such symptoms are very common among professionals who make considerable efforts to achieve high performance, too little physical stress can, to a certain extent, indicate the presence of understimulation at work. In such a situation, staff may perform poorly in terms of accomplishing task goals on time and within budget and in cooperating with other colleagues, all of which will further reduce job satisfaction and organisational commitment.

At the same time, an inverted U-shaped relationship was also found between *emotional stress* and performance. Construction personnel suffering from too much emotional stress will experience problems like depression, frustration, anxiety and loss of temper. This will directly result in performance impairment, influence their interaction with colleagues, reduce their sense of belonging and ultimately lead to an intention to leave the company. On the other hand, too little emotional stress may cause boredom and a feeling of being insufficiently challenged at work, which subsequently reduces interest and motivation and impairs performance.

7.1.4   Coping Behaviours of Construction Personnel

Since stress is not necessarily bad, the aim of stress management should not be to eliminate it entirely but instead to manage it properly, so that performance can be optimised (Haynes and Love 2004; Leung, Liu and Wong 2006). When suffering from stress, construction personnel cope by making cognitive and behavioural efforts to balance their environment, internal demands and emotions (Djebarni 1996). *Coping behaviours* can be classified into three categories: problem-, emotion- and meaning-based coping (Lazarus and Folkman 1984) (see Chapter 6). The findings of the empirical study show a total of 15 coping behaviours adopted by construction personnel, including problem-based (planful problem solving, instrumental support seeking, confrontive coping and positive reappraisal: Djebarni 1996; Greenglass, Schwarzer and Taubert 1999; Leung, Liu and Wong 2006);

emotion-based (emotional support seeking, emotional discharge, escapism and self-control: Folkman, Schaefer and Lazarus 1979; Pearlin and Schooler 1978; Schafer 2000) and meaning-based coping (present focus, acceptance, non-judgmental, describing, awareness, non-reactivity and beginner's mind: Grossman et al. 2004; Kabat-Zinn et al. 1992).

Among all these coping behaviours, construction personnel were most likely to use problem-based and the least likely to use emotion-based coping behaviours. *Those aged 29 or below preferred to adopt escapism and beginner's mind and avoided planful problem solving, instrumental support seeking, present focus and awareness.* More experienced construction personnel will usually be able to handle project problems and maintain higher awareness of the present than their younger colleagues. Since younger personnel have only recently begun working in the construction industry, they may not have amassed enough experience or ability to solve the problems encountered in their daily work, nor established a network of relationships from which to seek instrumental support. Therefore, younger personnel may find it easier to remain curious and apply beginner's mind to their workplace and tasks. However, if they encounter too many problems and find they cannot cope, they may choose to escape from them rather than confront the issues.

Male construction personnel tended to adopt more confrontive coping, while women preferred emotional support seeking. Since men tend to engage in direct action more often than women and are less likely to avoid confrontation (e.g., Vingerhoets and Van Heck 1990), this is not surprising. On the other hand, men and women are socialised in different ways insofar as men are encouraged to rely on themselves, while women are encouraged to seek support from others (Stokes and Wilson 1984). Again, therefore, it is not surprising that the women in this study were more likely to adopt emotional support seeking than their male colleagues.

A further Pearson correlation analysis was conducted to investigate the relationships between stress, coping behaviours and performance for construction personnel. The results indicated that: (i) confrontive coping, emotional discharge, escapism and self-control were positively related to stress and instrumental support seeking, present focus, non-judgmental and awareness were negatively associated with it; (ii) four problem-based coping behaviours (planful problem solving, instrumental support seeking, confrontive coping and positive reappraisal) were positively related to most types of performance; (iii) among the emotion-based coping behaviours, only emotional discharge was related to performance; and (iv) self-control, acceptance, describing, awareness and beginner's mind were positively related to performance.

Construction projects normally involve complicated problems. Hence, *problem-based coping behaviours* can be employed to *solve these problems, seek support and positively reappraise* the stressors. These behaviours can enable personnel to deal with problems efficiently and positively, achieve better task performance and, subsequently, develop closer relationships with colleagues and establish a sense of belonging to the company (Oakland and Ostell 1996). Those using instrumental support are less likely to worry about problems and to suffer from headaches, sleep disorders and back pain. However, *confrontive coping* has a positive relationship with both emotional and physical stress. Construction personnel using this approach will express their anger to others more often and approach their tasks aggressively. They are more likely to suffer from task demand and have poor relationships with others. The more confrontive coping is used, the more emotional and physical stress will be experienced.

Construction personnel adopting *emotion-based coping behaviours* (emotional discharge, escapism and self-control) will be able to deal with their negative feelings. However, the underlying problems will remain unsolved and the work stress will not be managed. *Escapism* is not a sustainable response to work stress. Personnel who control their emotions and feelings over a prolonged period of stress are more likely to suffer from serious emotional stress and damage their physical health (Violanti 1993). Moreover, *emotional discharge* also leads to physical problems such as headaches, stomach aches and coughing. Construction personnel who take this approach will not be able to handle problems with the project or improve their performance, resulting in low levels of personal satisfaction.

Among the *meaning-based coping behaviours*, construction personnel who remain focused on the present situation may find it easier to reduce their stress levels. Being *non-judgmental* cannot solve problems or improve performance, but it will lead to decreased physical stress. If personnel describe their problems and think them through using beginner's mind, it will be easier for them to evaluate the situation and begin to derive solutions from new perspectives. Hence, using *describing and beginner's mind*, construction personnel can solve their problems more efficiently, improve task performance and maintain good relationships with colleagues.

7.1.5   An Integrated Stressor–Stress–Coping Behaviours– Performance Model

Of the 21 stressors studied here, 11 were shown to have a significant impact on the three kinds of stress (work, physical and emotional). Construction personnel with Type A behaviour who encounter work–home conflict and poor interpersonal relationships are more likely to suffer from stress. At the same time, task stressors (including work overload, role ambiguity, role conflict, effort–reward imbalance and lack of feedback from superiors) can also trigger stress. In a construction organisation, stress can be induced by lack of manpower support, while the physical stressors of a poor office environment and unsafe site environment may also be relevant factors in daily working life. On the other hand, distrust among colleagues, organisational centralisation and lack of career support will induce emotional stress and work underload also causes physical stress. A poor site environment is directly related to emotional and physical stress, while work stress is induced by a poor home life.

All three kinds of stress were also related to performance in that there are both negative and inverted U-shaped relationships between them. Too much or too little stress leads to making the wrong decisions, poor task performance, impaired interpersonal relationships, unsafe behaviours and a greater intention to leave the company. Only a moderate level of work, emotional and physical stress will stimulate individual interest in and motivation for work, improve efficiency, facilitate cooperation with colleagues and foster commitment and a sense of belonging. However, although physical stress also has an inverted U-shaped relationship with performance, it should still be reduced because prolonged exposure to it is associated with a greater risk of morbidity and mortality (Nixon et al. 2011).

Problem-based coping behaviours (planful problem solving, instrumental support seeking, confrontive coping and positive reappraisal) are more useful in dealing with problems and improving ultimate performance. Instrumental support seeking also helps to relieve physical stress, while confrontive coping causes feelings of anger and aggression and triggers emotional and physical stress. The adoption of emotion-based coping (emotional discharge, escapism and self-control) increases stress levels because such approaches have no impact on solving the problems which are causing the stress. In addition, emotional discharge also decreases personal performance. Moreover, the stress of construction personnel can be reduced if they focus on the present situation and maintain a non-judgmental attitude and awareness. Those who cope with stress using describing and beginner's mind are more likely to perform well, while the use of non-judgmental behaviours tends to decrease personal satisfaction and sense of belonging for construction personnel.

7.2   Practical Recommendations

The integrated model presented in this book gives rise to a number of practical recommendations for optimising the performance of construction personnel through effective stress management. To enhance personal satisfaction, interpersonal relationships, project outcomes, sense of belonging and intention to stay in the construction industry, work and emotional stress should be maintained at a moderate level and physical stress should be reduced. To achieve this, it is suggested that the focus should be on managing stressors (personal, interpersonal, task, organisational and physical) and equipping staff with coping skills (mainly problem- and meaning-based coping).

7.2.1   Recommendations for Managing Stressors

As Type A behaviour and work–home conflict induce all three kinds of stress, construction companies should pay particular attention to them. In order to understand the personality traits of construction personnel (such as Type A behaviour), *relevant behavioural and psychological tests should be conducted regularly*. It will also be helpful to create an organisational culture which *discourages overtime* and *encourages optimised productivity during office hours*. On the other hand, it is also essential for construction personnel to learn how to *balance work and home life* (Leung, Chan, Chong and Sham, 2008). If possible, employers could *invite the family members of construction personnel to participate in informal social gatherings* outside office hours such as Christmas functions and sporting competitions (Leung et al. 2009).

Since construction projects involve multiple parties, poor interpersonal relationships are among the most important stressors. In order to improve interpersonal relationships, *training in communication skills and team-building techniques* should be offered. Team-building activities are strongly recommended as they provide opportunities for personnel to work with colleagues and develop trust in each other (Leung, Sham and Chan 2007).

Both work over- and underload induce stress. Construction companies should therefore consider *regular progress meetings to review the workload, work progress and difficulties* faced by construction personnel. Due to the dynamic nature of construction projects, workload should be reviewed regularly for each individual. Regular progress meetings could also be held so that construction personnel can provide feedback and opinions about their work allocations. As younger personnel in this study experienced work underload, senior managers are encouraged to *delegate appropriate tasks to their subordinates,* including both simple and more complicated jobs depending on ability. To avoid role ambiguity and role conflict, it is necessary for construction companies and leaders to specify the job responsibilities allocated to their staff and provide relevant training for younger personnel. Moreover, firms should *establish a stable performance evaluation system and provide adequate welfare support* in order to help staff strike a balance between effort and reward. They could also recognise outstanding performance through tangible or intangible rewards. On the other hand, adequate safety equipment, in terms of both quantity and quality, should be provided on site to ensure the safety as well as the emotional and physical health of construction personnel.

Organisational centralisation disempowers construction personnel and results in emotional stress. Apart from the specification of job and positional responsibility, it is also suggested that companies *increase the opportunity for construction personnel at different management levels to participate in making decisions of various levels of importance* and *to influence processes.* To address this organisational stressor, it is also important to provide sufficient career and manpower support for employees. Senior *personnel can also share their invaluable experience* with subordinates and give them advice and feedback.

Physical factors also significantly influence the stress level of construction personnel, so a clean, tidy and safe environment, whether on site or in the office, is essential. To provide a comfortable working environment, construction companies should *ensure proper functioning of the indoor air ventilation and temperature systems* so as to maintain constant temperature. Office design should also take *lighting*, both artificial and natural, and *working space*, both size and privacy, into consideration (Leung et al. 2009). Since too much noise can negatively affect work performance, acoustic insulation wall and ceiling panels are suggested to minimise noise, particularly for those who work on site (Leung et al. 2005). Firms should also choose machines with low noise and vibration when purchasing plant and equipment such as excavators, concrete mixers and electric drills. If necessary, personnel on site need to wear earmuffs to avoid hearing damage, masks to minimise the influence of dust and helmets to protect the head. To decrease the risk of heatstroke, it is suggested that companies *stagger working hours* (such as early morning and late at night) in periods of hot weather (Morioka, Miyai and Miyashita 2006) and *spray water on the site to decrease dust levels during dry weather.*

7.2.2   Recommendations for Managing Stress

To optimise the impact of work stress, construction stakeholders should not simply aim to eliminate it at the individual level, but keep reviewing and monitoring the abilities of their construction personnel and allocate suitable job responsibilities and workload accordingly (Leung, Chan and Olomolaiye 2008). It would also be helpful to *provide assistance and support to construction personnel, especially younger ones*, firstly to improve *safety behaviours* (by means such as rewarding safe behaviours and punishing unsafe behaviours) and secondly to improve project outcomes and personal satisfaction.

Emotional stress should be managed so as to improve interpersonal relationships, project outcomes and sense of belonging as well as intention to stay. *Stress management workshops* (along the lines of the mental health first aid workshops organised regularly for university students) or *counselling meetings* would be good options for construction organisations seeking to promote the acquisition of various relevant skills (Gates 2003). It is also suggested that firms should *provide periodic all-round stress appraisals* for staff to share their feelings, provide mutual emotional support and understand both the organisational value of what they do and their expected and actual abilities.

To maintain a healthy body, the practise of regular fitness exercises every morning is also strongly recommended. Construction organisations could arrange *events to promote physical fitness* (such as sports days, marathon runs) or allocate a small space in their offices for light *gym equipment* such as running, spinning and cycling machines. As well as professional practical knowledge, it is important to conduct *seminars and workshops* (such as training in use of the fitness equipment, healthcare seminars) to enhance knowledge among construction personnel of how to promote physical wellbeing. Furthermore, employers could offer annual *physical checkups* to their staff. It is better to rectify any potential or minor physical stress (in terms of normal wear and tear on the body) in advance rather than to try to cure serious physical issues such as headaches and back pain.

7.2.3   Recommendations for Coping with Stress

Based on the results of previous studies, it is strongly recommended that construction personnel adopt the three problem-based (planful problem solving, instrumental support seeking and positive reappraisal) and three meaning-based coping strategies (describing, awareness and beginner's mind) to manage their inner attitudes when confronting difficulties and stressful situations. Other coping behaviours may have unwanted side effects for construction personnel, such as increasing their stress level and diminishing performance. For example, emotional discharge not only exacerbates physical stress levels, but also reduces personal satisfaction. Confrontive coping improves personal satisfaction and project outcomes, but can also induce stress. Non-judgmental coping strategies reduce stress, but construction personnel may not be able to complete their tasks by using this alone.

It is also suggested that construction personnel try to solve problems directly through better planning and utilisation of instrumental support. Various workshops and activities for supporting problem solving are already in use in the industry, including value management, partnering and team building. *Partnering workshops are normally conducted* at the beginning of construction projects in order to understand the requirements of all parties and seek appropriate support. When facing difficult tasks, construction personnel are encouraged to *reappraise the problems and treat difficulties as challenges and opportunities* for personal development. If necessary, they can also adopt value management techniques to review problems via a systematic team decision-making process. *Team-building activities* can be used to improve team spirit among colleagues and provide instrumental support.

When coping with stress, construction personnel should focus on the present situation, describe the problem and think over the solutions with curiosity and beginner's mind. To facilitate these meaning-based coping behaviours, a *mindfulness-based stress reduction (MBSR) workshop* is recommended. Moreover, *informal mindfulness practices*, such as mindful eating, walking and breathing, can help to cultivate the inner resources required to handle any unexpectedly complicated or difficult tasks arising in daily working life. Due to the side effects of emotion-based coping, construction personnel should try to avoid reliance on emotional discharge, escapism and self-control alone.

Unfortunately, there are various real-life examples showing the prominent and negative impact of stress on younger construction personnel, such as the intention to leave the industry and the threat or even the actual act of committing suicide. The emergence of these recent cases suggests, to a certain extent, that people in our industry are suffering from more stress than ever before (which is consistent with the findings of a survey conducted by the CIOB [2006]) and that it is having a particular impact on early-career construction personnel who have not yet developed sufficient adaptive stress management skills. Hence, a stress monitoring system and management training are very important for firms and should be provided to younger personnel in particular. However, construction companies are not the only stakeholders who can organise stress management activities. All construction institutions can make an impact on creating an industry-wide stress management culture. Indeed, some are already taking the lead in organising stress management seminars and workshops for personnel in the sector (Weinberg, Sutherland and Cooper 2010). In order to nurture the long-term resilience of all involved in the industry, stress management work should also be incorporated as part of professional training in universities or tertiary education institutions. All in all, training in stress management should be provided not only to younger construction staff, but also to those who will be the construction personnel of the future; that is, students.

7.2.4   Summary of Practical Recommendations

In order to optimise performance, construction personnel, especially younger personnel, should manage their *stressors* and *stress* and learn to *cope* with stress effectively in work and life. Our practical recommendations for stress management to optimise the performance of construction personnel are summarised in Table 7.1.

7.3   Recommendations for Further Research

7.3.1   Triangulation and Longitudinal Studies

Although more and more research studies are focusing on stress management in the construction sector, the majority adopt a quantitative approach by measuring the stress levels of construction personnel using a questionnaire and applying various statistical tools to associate stress level with various variables such as stressors and performance (see e.g. Bowen et al. 2013; Love and Edwards 2005; Leung, Chan and Olomolaiye 2008). Among the various statistical tools, structural equation modelling is considered the most comprehensive approach to deriving interdependent relationships. However, basing such a model on cross-sectional data may lead to uncertainty about causation, because causality and the directions of path relationships have been hypothesised by the researcher. Therefore, to enhance the extent to which we can have confidence in the results of stress management research, two approaches are recommended: (i) structural equation models should be built based on a conceptual research model supported by sound theory and based on an extensive literature review, preferably supported by the results of a preliminary focus group study (see e.g. Chan, Leung and Yu 2012; Leung, Chan and Yu 2012); and (ii) qualitative research, such as a case study, should be conducted to cross-validate the model. If any paths in the model demonstrate causal directions inconsistent with the case study, this should be highlighted and discussed in the results. However, such mixed-method approach can be carried out on a single occasion and essentially represents a snapshot of what a particular group of personnel faced in their stressful work environment at a given time. In order to validate causal relationships by observing and tracking changes in the stress management processes of construction personnel over time, we recommend a longitudinal study which collects repeat data using interviews and surveys over an extended period (such as four months). A good example of this would be an evaluation of the effectiveness of a particular stress management intervention. By measuring stress levels and related dimensions (such as stressors and performance)

Table 7.1   Recommendations for optimising stress and performance of construction personnel

In order to	Suggestions for management of stressors
Understand Type A behaviour	Undertake behavioural and psychological tests to understand the personality of construction personnel and their stress experiences
Handle work–home conflict	Build an organisational culture which discourages work overtime, while encouraging optimised productivity during office hours
	Encourage family members of construction personnel to participate in informal social events
Improve interpersonal relationships	Organise team-building activities
	Provide training on communication skills
Cope with work overload/ underload	Hold regular progress meetings and review workload
	Encourage senior construction personnel to delegate tasks to subordinates
Deal with role ambiguity and conflict	Specify the responsibilities of construction personnel clearly
Avoid reward–effort imbalance	Establish clear performance evaluation systems
	Provide adequate welfare support
	Recognise outstanding performance with tangible or intangible rewards
Provide adequate safety equipment	Provide adequate and appropriate safety equipment to reduce the possibility of accidents and injuries
Provide feedback from superiors	Provide feedback or advice for subordinates from senior personnel and management
Reduce organisational centralisation	Increase the opportunities for construction personnel at different management levels to participate in different types and levels of decision making
Provide manpower support	Allocate tasks appropriately and ensure sufficient manpower is available
Improve office environment	Ensure proper functioning of the indoor air ventilation and temperature systems
	Take into account lighting, both artificial and natural, and work space, both size and privacy, when designing offices
Improve site environment	Select procedures and materials on the basis of the well-being of on-site personnel
	Stagger work hours in hot weather, provide masks and spray water on the construction site
In order to	**Suggestions for management of stress**
Optimise work stress	Regularly review the workload and expectations of construction personnel and reallocate workload at regular intervals
	Provide assistance and support about safety, particularly to younger personnel
Reduce physical stress	Organise events to promote physical fitness (such as sports days and marathons) and provide fitness equipment in the workplace
	Conduct regular seminars and workshops for physical health management
	Provide annual physical checkups
Optimise emotional stress	Offer stress management workshops, counselling services and periodic stress appraisals

(Continued)

Table 7.1 *(Continued)*

In order to	Suggestions for management of stressors
In order to	**Suggestions for encouraging staff to cope with stress**
Plan and establish a network for instrumental support	Conduct partnering workshops at the start of construction projects, reappraise problems, treat difficulties as challenges and opportunities, conduct value management workshops for solving problems as a team and organise team-building activities to develop team spirit
Conduct mindfulness practices	Offer MBSR workshops and encourage informal mindfulness practices such as mindful eating, mindful walking and body scan
Avoid inappropriate coping behaviours	Avoid maladaptive coping behaviours such as confrontive coping, emotional discharge, escapism, self-control and non-judgmental behaviours

pre- and post- intervention, effectiveness can be assessed, which in turn will facilitate the enhancement and redesign of such training materials.

7.3.2   Physiological Stress Measurement

As mentioned above, the majority of stress management studies in the construction sector are survey-based, supplemented by limited qualitative studies. The methods adopted in these studies to measure stress are generally subjective, such as self-administered questionnaires and self-descriptions of stress experiences in interviews. In order to cross-validate such subjective evaluations, objective physiological measurements (i.e. biological indicators of stress, such as blood pressure and cortisol levels) are also recommended. Blood pressure has been confirmed as one of the more useful physiological indicators of individual stress levels (Fried et al. 1984; Melin et al. 1999). In addition, the level of cortisol in saliva has been found to be sensitive to a wide range of chronic and acute stressors (Tse and Bond 2004), such as job demand (Kunz-Ebrecht et al. 2004; Lundberg and Hellström 2002), financial strain (Steptoe et al. 2005), military training (Clow et al. 2006), and laboratory psychosocial stressors (Dickerson and Kemeny 2004). In addition, salivary cortisol is also affected by positive psychological stress (Chan et al. 2006; Lai et al. 2005). This would be an essential, significant and important step for the development of stress management research in the construction sector.

7.3.3   Cross-Cultural Stress Management

It is well known that the construction sector is becoming more and more globalised. Considerable numbers of construction firms are expanding their business to international markets, meaning that they need to relocate existing personnel and/or employ new personnel to work as expatriates.

Spend on expatriate managers is probably "the single largest expenditure most companies make on any one individual except for the chief executive officer" (Black and Gregersen 1999: 53). The roles of expatriates, such as knowledge transfer and management and organisational development, are essential to firms (Edstrom and Galbraith 1977; Harzing 2001; Sparrow et al. 2004). However, expatriation does not necessarily guarantee success. Distress arising from expatriate assignments can lead to placement failure (Black and Gregersen, 1990; Brown 2008; Littrell et al. 2006), in terms of early return, failure of assignment and leaving the organisation altogether (Littrell et al. 2006). In fact, the failure rate has been estimated to range from 3% to as high as 70% of the expatriate population in different countries (Borstorff et al. 1997; Forster 1997; Lorange 2003). The authors have conducted other studies about the stressors, stress and coping behaviours of expatriate construction personnel working in mainland China, which has a different cultural background from Hong Kong (see e.g. Chan, Leung and Yu 2012; Leung, Chan and Yu 2012). However, there is still no comprehensive stress management model for expatriate construction personnel, who may adopt different types of coping behaviours and access different forms of organisational support due to the special stressors facing them (such as cultural orientation training to address the differences in approach between local and expatriate construction personnel). Further investigation of the stress management of expatriate construction personnel is recommended.

7.4   Conclusion

This chapter has summarised the major research results presented in this book and set out the integrated stressor–stress–coping behaviours–performance model. A total of 21 stressors have been identified, including the personal (Type A behaviour and work–home conflict); interpersonal (poor interpersonal relationships and distrust); task (work overload/underload, role ambiguity and conflict, lack of job autonomy, effort–reward imbalance, inadequate safety equipment and lack of feedback from superiors); organisational (organisational centralisation and formalisation, lack of career support, lack of financial support and lack of manpower) and physical (poor office, site and home environment and unsafe site environment). Among all these, 11 stressors (Type A behaviour, work–home conflict, poor interpersonal relationships, work overload, role ambiguity/conflict, effort–reward imbalance, lack of manpower support and feedback from superiors, poor office environment and unsafe site environment) have been shown to have a significant impact on the three kinds of stress (work, physical and emotional). Distrust among construction personnel, organisational centralisation and lack of career support induce emotional stress, while work underload also causes physical stress. Poor site environment has a direct influence on emotional and physical stress and a poor home environment leads to work stress.

The consequences of stress for construction personnel have been identified in terms of personal satisfaction, interpersonal relationships, project outcomes, safety behaviours, sense of belonging and intention to stay with the company. All three kinds of stress are related to these types of performance. Both work and emotional stress have inverted U-shaped relationships with performance: work stress with personal satisfaction and safety behaviours and emotional stress with poor interpersonal relationships, project outcomes, sense of belonging and intention to stay. Although the physical stress of construction personnel also has an inverted U-shaped relationship with some performance factors, it is nevertheless recommended to reduce physical stress because prolonged exposure can induce further, serious health problems (Nixon et al. 2011).

Construction personnel adopt various coping behaviours, which include problem- (planful problem solving, positive reappraisal, confrontive coping. and instrumental support seeking), emotion- (emotional support seeking, emotional discharge, escapism and self-controlling) and meaning-based (present focus, acceptance, non-judgmental, describing, awareness, non-reactivity and beginner's mind) coping behaviours. Problem-based coping behaviour has a positive impact on performance. Instrumental support seeking also helps to relieve physical stress, while confrontive coping causes personnel to feel angry and aggressive and triggers emotional and physical stress. However, the adoption of emotion-based coping increases stress levels and has less impact on performance, because it does nothing to resolve the problems the individual is facing. Moreover, construction personnel also use meaning-based coping behaviours to cope with stress. Stress can be reduced by focusing on the present situation and maintaining a non-judgmental attitude and awareness. Moreover, performance can also be improved by using describing and beginner's mind. In conclusion, stress management, based on an understanding of stressors, stress and coping behaviours, plays an essential role in improving the performance of personnel involved in complicated industrial projects.

Lastly, three recommendations for further academic research have been made. Firstly, in view of the dominance of quantitative approaches, we recommend adopting a triangulation methodology (i.e. incorporating both qualitative and quantitative approaches to cross-validate results) and a longitudinal approach (such as the study of the effectiveness of a stress management intervention). Secondly, since the measurement methods adopted in the stress management research in the construction sector are mainly subjective (such as self-administered surveys and self-descriptions in interviews), we recommend the use of objective physiological measurements of stress (such as levels of salivary cortisol). Thirdly, due to the international trend in the global construction sector, it is inevitable that some construction personnel will work as, or with, expatriates from other countries who carry different cultural values. Further studies are recommended to investigate and develop an integrated model of the stress management process for expatriate construction personnel.

References

Andrews, R., Boyne, G.A., Law, J. and Walker, R.M. (2009) Centralization, organizational strategy, and public service performance. *Journal of Public Administration Research and Theory*, 19(1), 57–80.

Beehr, T. A. (1995) *Psychological Stress in the Workplace*. London: Routledge.

Beehr, T.A. and Jex, S.M. (2001) The management of occupational stress. In C.M. Johnson, W.K. Redmon and T.C. Mawhinney (eds), *Handbook of Organizational Performance*. New York: Haworth Press.

Black, J.S. and Gregersen, H.B. (1990) Expectations, satisfaction and intention to leave of American expatriate managers in Japan. *International Journal of Intercultural Relations*, 14(4), 485–506.

Black, J.S. and Gregersen, H.B. (1999) The right way to manage expats. *Harvard Business Review*, 77(2), 52–62.

Borstorff, P.S., Harris, S.G., Field, H.S. and Giles, W.F. (1997) Who'll go? A review of factors associated with employee willingness to work overseas. *Human Resource Planning*, 20(3), 29–40.

Bowen, P., Edwards, P. and Lingard, H. (2013) Workplace stress among construction professionals in South Africa: the role of harassment and discrimination. *Engineering, Construction and Architectural Management*, 20(6), 620-635.

Brown, R.J. (2008) Dominant stressors on expatriate couples during international assignments. *International Journal of Human Resource Management*, 19(6), 1018–1034.

Cannon, W.B. (1914) The interrelations of emotions as suggested by recent physiological researches. *American Journal of Psychology*, 25(2), 256–282.

Cannon, W.B. (1927) Bodily changes in pain, hunger, fear and rage. *Southern Medical Journal*, 22(9), 870.

Caplan, R.D. and Jones, K.W. (1975) Effects of work load, role ambiguity and Type-A personality on anxiety, depression and heart rate. *Journal of Applied Psychology*, 60, 713–719.

Chan, C.L.W., Tso, I.F., Ho, R.T.H., Ng, S.M., Chan, C.H.Y., Chan, J.C.N., Lai, J.C.L. and Evans, P.D. (2006) The effect of a 1-hour Eastern stress management session on salivary cortisol. *Stress and Health*, 22, 45–49.

Chan, I.Y.S., Leung, M.Y. and Yu, S.W. (2012) Managing stress of Hong Kong expatriate construction professionals in Mainland China: A focus group study to exploring individual coping strategies and organizational support. *Journal of Construction Engineering and Management*, 138(10), 1150–1160.

Chrousos, G.P. and Gold, P.W. (1992) The concepts of stress and stress system disorders: Overview of physical and behavioral homeostasis. *Journal of American Medical Association*, 267(9), 1244–1252.

CIOB (2006) Occupational stress in the construction industry, CIOB Published National Stress Survey *Results*, retrieved http://www.ciob.org.uk/resources/publications on 21 June 2008.

Clow, A., Edwards, S., Owen, G., Evans, G., Evans, P., Hucklebridge, F. and Casey, A. (2006) Post-awakening cortisol secretion during basic military training. *International Journal of Psychophysiology*, 60(1), 88–94.

Cooper, C.L. (2001) *Organization Stress: A Review and Critique of Theory, Research and Application*. Thousand Oaks, CA: Sage.

Cox, T. (1993) Stress research and stress management: Putting theory to work. *HSE Contract Research Report*, No. 61/1993.

Daniels, K. and Guppy, A. (1994) Occupational stress, social support, job control, and psychological well-being. *Human Relations*, 47(12), 1523–1544.

Dickerson, S.S. and Kemeny, M.E. (2004) Acute stressors and cortisol responses: A theoretical integration and synthesis of laboratory research. *Psychological Bulletin*, 130(3), 355–391.

Djebarni, R. (1996) The impact of stress in site management effectiveness. *Construction Management and Economics*, 14(4), 281–293.

Djebarni, R. and Lansley, R. (1995) Impact of site managers' leadership on project effectiveness. *Proceedings of the 1st International Conference on Construction Project Management*, Singapore, 123–131.

Driskell, J.E. and Salas, E. (1991) Overcoming the effects of stress on military performance: Human factors, training, and selection strategies. In R. Gal and A. Mangelsdorff (eds), *Handbook of Military Psychology*. Chichester: John Wiley, 183–193.

Edstrom, A. and Galbraith, J.R. (1977) Transfer of managers as a coordination and control strategy in multinational organization. *Administrative Science Quarterly*, 22, 248–263.

Eisenberger, R., Huntington, R., Hutchison, S. and Sowa, D. (1986) Perceived organizational support. *Journal of Applied Psychology*, 71, 500–7.

Emslie, C., Hunt, K. and Macintyre, S. (2004) Gender, work–home conflict and morbidity amongst white-collar bank employees in the UK. *International Journal of Behavioral Medicine*, 11(3), 127–134.

Folkman, S., Schaefer, C. and Lazarus, R.S. (1979) Cognitive processes as a mediator of stress and coping. In V. Hamilton and D.M. Warburton (eds), *Human Stress and Cognition: An Information Processing Approach*. London: Wiley.

Forster, N. (1997) The persistent myth of high expatriate failure rates: A reappraisal. *International Journal of Human Resource Management*, 8(4), 414–433.

Francesco, A. and Gold, B. (2005) *International Organizational Behavior*. New Jersey: Pearson Prentice Hall.

Fried, Y., Rowland, K.M. and Ferris, G.R. (1984) The physiological measurement of work stress: a critique. *Personnel Psychology*, 37(4), 583-615.

Ganster, D.C. and Rosen, C.C. (2013) Work stress and employee health: A multidisciplinary review. *Journal of Management*, 39(5), 1085–1117.

Gates, E. (2003) Workplace stress counselling, *Occupational Safety and Health*, 33(5), 40-44.

Gmelch, W.H. (1982) *Beyond Stress to Effective Management*. New York: Wiley.

Greenglass, E., Schwarzer, R. and Taubert, S. (1999) The Proactive Coping Inventory (PCI): A multidimensional research instrument. *The 20th International Conference of the Stress and Anxiety Research Society*. Cracow, Poland.

Greenhaus, J.H. and Parasuraman, S. (1987) A work–nonwork interactive perspective of stress and its consequences. In J.M. Ivancevich and D.C. Ganster (eds), *Job Stress: From Theory to Suggestion*. New York: Haworth.

Grossman, P., Niemann, L., Schmidt, S. and Walach, H. (2004) Mindfulness-based stress reduction and health benefits-A meta-analysis. *Journal of Psychosomatic Research*, 57(1), 35–44.

Hancock, P.A. and Warm, J.S. (1989) A dynamic model of stress and sustained attention. *Human Factors*, 31(5), 519–537.Harzing, A.W. (2001) Of bears, bumblebees, and spiders: the role of expatriates in controlling foreign subsidiaries, *Journal of World Business*, 36(4), 366-379.

Haynes, N.S. and Love, P.E.D. (2004) Psychological adjustment and coping among construction project managers. *Construction Management and Economics*, 22, 129–140.

Health and Safety Executive of the UK Government (2006) *Stress-related and psychological disorders*. Retrieved from http://www.hse.gov.uk/statistics/causdis/stress.htm.

Hockey, G.R.J. (1997) Compensatory control in the regulation of human performance under stress and high workload: A cognitive–energetical framework. *Biological Psychology*, 45(1), 73–93.

Hockey, G.R.J. (1993) Cognitive-energetical control mechanisms in the management of work demands and psychological health. In A.D. Baddeley and L. Weiskrantz (eds). *Attention, Selection. Awareness and Control: A Tribute to Donald Broadbent.* Oxford: Oxford University Press.

Kabat-Zinn, J., Massion, A.O., Kristeller, J., Peterson, L.G., Fletcher, K.E. and Pbert, L. (1992) Effectiveness of a meditation-based stress reduction program in the treatment of anxiety disorders. *American Journal of Psychiatry*, 149, 936–943.

Keenan, A. and Newton, T.J. (1987) Work difficulties and stress in young professional engineers. *Journal of Occupational Psychology*, 60(2), 133–145.

Kunz-Ebrecht, S.R., Kirschbaum, C., Marmot, M. and Steptoe, A. (2004) Differences in cortisol awakening response on work days and weekends in women and men from the Whitehall II cohort. *Psychoneuroendocrinology*, 29(4), 516–528.

Lai, J.C.L., Evans, P., Ng, S.H., Chong, A.M.L., Siu, O.T., Chan, C.L.W., Ho, S.Y.M., Ho, R.T.H., Chan, P. and Chan, C.C. (2005) Optimism, positive affectivity, and salivary cortisol. *British Journal of Health Psychology*, 10, 467–484.

Langeland, W. and Olff, M. (2008) Psychobiology of posttraumatic stress disorder in pediatric injury patients: A review of literature. *Neuroscience and Biobehavioral Reviews*, 32(1), 161–174.

Lazarus, R.S. and Folkman, S. (1984) *Stress, Coping and Adaptation.* New York: Springer-Verlag.

Leung, M.Y., Olomolaiye, P., Chong, A. and Lam, C.C. (2005) Impacts of stress on estimation performance in Hong Kong. *Construction Management and Economics*, 23(9), 891–903.

Leung, M.Y., Liu, A.M.M. and Wong, M.K. (2006) Impact of stress-coping behaviors on estimation performance in Hong Kong. *Construction Management and Economics*, 24(1), 56–57.

Leung, M.Y., Sham, J. and Chan, Y.S. (2007) Adjusting stressors – Job-demand stress in preventing rustout/burnout in estimators. *Surveying and Built Environment*, 18(1), 17–26.

Leung, M.Y., Chan, Y.S., Chong, A. and Sham, J.F.C. (2008) Developing structural integrated stressor–stress models for clients' and contractors' cost engineers. *Journal of Construction Engineering and Management*, 134(8), 635–643.

Leung, M.Y., Chan, Y.S. and Olomolaiye, P. (2008) Impact of stress on the performance of construction project managers. *Journal of Construction Engineering and Management*, 134(8), 644–652.

Leung M.Y., Chan Y.S., Yu J.Y. (2009) Integrated model for the stressors and stresses of construction project managers. *Construction Engineering and Management,* ASCE, 135(2), 126–134.

Leung, M.Y., Chan, Y.S. and Yuen, K.W. (2010) Impacts of stressors and stress on the injury incidents of construction workers in Hong Kong. *Journal of Construction Engineering and Management*, 136(10), 1093–1103.

Leung, M.Y., Chan, Y.S.I. and Dongyu, C. (2011).Structural linear relationships between job stress, burnout, physiological stress and performance of construction project managers. *Engineering, Construction and Architectural Management*, 18(3), 312–328.

Leung, M.Y., Chan, Y.S. and Yu, J.Y. (2012) Preventing construction worker injury incidents through the management of personal stress and organizational stressors. *Accident Analysis and Prevention*, 156–166.

Lazarus, R.S. and Folkman, S. (1984) *Stress, Appraisal and Coping.* New York: Springer.

Littrell, L.N., Salas, E., Hess, K.P., Paley, M. and Riedel, S. (2006) Expatriate preparation: A critical analysis of 25 years of cross-cultural training research. *Human Resource Development Review*, 5(3), 355–388.

Loosemore, M. and Waters, T. (2004) Gender differences in occupational stress among professionals in the construction industry. *Journal of Management in Engineering*, 20(3), 126–132.

Lorange, P. (2003) Developing global leaders. *BizEd*, 2(6), 24–28.

Love, P. and Edwards, D.J. (2005) Taking the pulse of UK construction project managers' health: Influence of job demands, job control and social support on psychological wellbeing. *Engineering, Construction and Architectural Management*, 12(1), 88–101.

Lundberg, U. and Hellström, B. (2002) Workload and morning salivary cortisol in women. *Work and Stress*, 16(4), 356–363.

Melin, B., Lundberg, U., Soderlund, J. and Granqvist, M. (1999) Psychological and physiological stress reactions of male and female assembly workers: A comparison between two different forms of work organization. *Journal of Organizational Behavior*, 20(1), 47–61.

Morioka, I., Miyai, N. and Miyashita, K. (2006) Hot environment and health problems of outdoor workers at a construction site. *Industrial Health*, 44(3), 474–480.

Mullen, J. (2004) Investigating factors that influence individual safety behavior at work. *Journal of Safety Research*, 35(3), 275–285.

Nixon, A.E., Mazzola, J.J., Bauer, J., Krueger, J.R. and Spector, P.E. (2011) Can work make you sick? A meta-analysis of the relationships between job stressors and physical symptoms. *Work and Stress*, 25(1), 1–22.

Nyssen, A.S, Hansez, I., Baele, P., Lamy, M. and Keyser V.D. (2003) Occupational stress and burnout in anaesthesia. *British Journal of Anaesthesia*, 90(3), 333–337.

Oakland, S. and Ostell, A. (1996) Measuring coping: A review and critique. *Human Relations*, 49(2), 133–155.

Parasuraman, R. and Rizzo, M. (2008) *Neuroergonomics: The Brain at Work*. Oxford: Oxford University Press.

Pearlin, L.I. (1989) The sociological study of stress. *Journal of Health and Social Behavior*, 30(3), 241–256.

Pearlin, L.I. and Schooler, C. (1978) The structure of coping. *Journal of Health and Social Behavior*, 2–21.

Puca, F., Genco, S., Prudenzano, M.P., Savarese, M., Bussone, G., Amico, D., Cerbo, R., ..., Marabini, S. (1999) Psychiatric comorbidity and psychosocial stress in patients with tension-type headache from headache centers in Italy. *Cephalalgia*, 19(3), 159–164.

Quick, J.C. and Quick, D.L. (1997) *Preventive Stress Management in Organizations*. Washington, DC: American Psychological Association.

Reichers, A.E, Wanous, J.P. and Steele, K. (1994) Design and implementation issues in socializing (and resocializing) employees. *Human Resource Planning*, 17, 17–25.

Schafer, W. (2000) *Stress Management for Wellness*. New York: Wadsworth.

Selye, H. (1964) *From Dream to Discovery*, New York: McGraw-Hill.

Selye, H. (1976) *Stress in Health and Disease*. Boston: Butterworths.

Sparrow, P., Brewster, C. and Harris, H. (2004) *Globalizing Human Resource Management*. London: Routledge.

Steptoe, A., Brydon, L. and Kunz-Ebrecht, S. (2005) Changes in financial strain over three years, ambulatory blood pressure, and cortisol responses to awakening. *Psychosomatic Medicine*, 67(2), 281–287.

Stokes, J.P. and Wilson, D.G. (1984) The inventory of socially supportive behaviors: dimensionality, prediction, and gender differences. *American Journal of Community Psychology*, 12(1), 53–69.

Teasdale, E.L. and Mckeown, S. (1994) Managing stress at work: The ICI-Zeneca Pharmaceuticals experience. In C.L. Cooper and S. Williams (eds), *Creating Healthy Work Organisations*. Chichester: John Wiley.

Tse, W.S. and Bond, A.J. (2004) Relationship between baseline cortisol, social functioning and depression: a mediation analysis. *Psychiatry Research*, 126(3), 197-201.

Vingerhoets, A.J. and Van Heck, G.L. (1990) Gender, coping and psychosomatic symptoms, *Psychological Medicine*, 29(1), 125-135.

Violanti, J.M. (1993) What does high stress police training teach recruits? An analysis of coping. *Journal of Criminal Justice*, 21(4), 411–417.

Weinberg, A., Sutherland, V. and Cooper, C.L. (2010) *Organizational Stress Management: A Strategic Approach*. Basingstoke: Palgrave Macmillan.

Yerkes, R.M. and Dodson, J.D. (1908) The relation of strength of stimulus to rapidity of habit formation. *Journal of Comparative Neurology and Psychology*, 18(5), 459–482.

Yip, B. (2008) Professional efficacy among building professionals. *The Conference for Building Professionals*. Gold Coast, Australia, 24–26.

Zaccaro, S.J. and Riley, A.W. (1987) Stress, coping and organizational effectiveness. *Occupational Stress and Organizational Effectiveness*, 1–28.

Index

Note: page numbers in italics refer to figures; page numbers in bold refer to tables.

Stress Management in the Construction Industry, First Edition.
Mei-yung Leung, Isabelle Yee Shan Chan and Cary L. Cooper.
© 2015 John Wiley & Sons, Ltd. Published 2015 by John Wiley & Sons, Ltd.

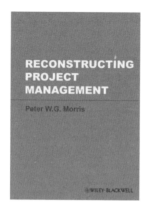
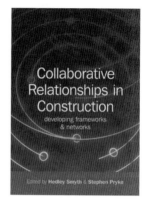
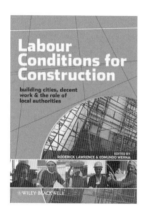
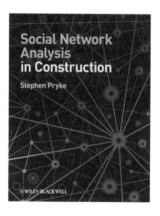
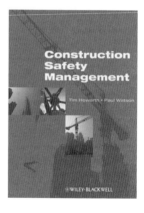

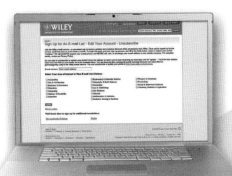